The Flower of Battle
MS M 383

The Flower of Battle
MS M 383

by Michael Chidester

With appendix by Jay Leccese

HEMA Bookshelf

A version of "Riddled in Ink: A Stylistic Comparison of Artwork in MS M 383 and Novati's Facsimile" was previously published on *Hroarr* on 24 October 2016.
http://hroarr.com/article/riddled-ink-stylistic-comparison-artwork-ms-m-383-novati-facsimile/

Manuscript scans of ms. M. 383 provided by the Morgan Library & Museum, New York.
Digital images of ms. Ludwig XV 13 courtesy of the J. Paul Getty Museum's Open Content Program.
Digital images of mss. Français 2258 and Latin 6069F courtesy of the Bibliothèque nationale de France, Paris
    Source: gallica.bnf.fr
Image of *Conquest of a Town* courtesy of the National Gallery of Art, Washington, D.C.
Scans of ms. Chart A.558 courtesy of the Forschungsbibliothek Gotha, Universität Erfurt
    https://dhb.thulb.uni-jena.de/receive/ufb_cbu_00009967 (CC-BY-SA 4.0)

Published by HEMA Bookshelf, LLC
411a Highland Ave #141
Somerville, MA, 02144
www.hemabookshelf.com

© 2021 HEMA Bookshelf.

All rights reserved.

No part of this work may be reproduced, distributed, or transmitted in any form or by any means, including photocopying, recording, or other electronic or mechanical methods, without the prior written permission of the author, except in the case of brief quotations embodied in critical reviews and certain other noncommercial uses permitted by copyright law.

Version 1.1, 2021

ISBN 978-1-953683-07-6

Library of Congress Control Number: 2021935616

Typeset in Libertinus Sans and Libertinus Serif Display, which are used under the Open Font License.
    http://libertine-fonts.org/
Titles in Waltari ML, used under license.

Cover design by Henry Snider based on photographs by Michael Chidester.

Printed by Lightning Source.

Table of Contents

Preface 1

Introduction 5

Dueling in the 15th century 11

    Galeazzo Cattaneo dei Grumelli – Giovannino da Baggio – Going to the woods

Codicological Description 23

    Materials and physical description – Dating and origin – Provenance – Content – Collation and foliation – Writing and decoration – Illustrations

Transcription and Translation 39

    Preface – Mounted spear – Mounted sword – Mounted grappling – Javelin vs horse – Armored spear – Armored sword – Sword in two hands – Special plays – Sword in one hand

Appendix: Riddled in Ink: A Stylistic Comparison of Artwork in MS M 383 and the Novati Facsimile 123

    Proposition – Relevance – Methodology – Stylistic analysis – Overall presentation – A retouched image in the Morgan – Forensic analysis – Handwriting – Title page – Discussion and conclusion – Suggestions for further research – About the author

Bibliography 151

About the Author 153

Acknowledgements

This book is the fruit of a decade and a half of study, and I was assisted by innumerable people along the way. I'll start by thanking the translators who gave freely of their time and expertise and whose labor helped me get my foot in the door: Hermes Michelini of the Knights of the Wild Rose, Matt Easton and Eleonora Litta of Schola Gladiatoria, and Rob Lovett and Mark Lancaster of the Exiles. Also Paolo Amassari and Brian Hunt, who did me the favor of producing additional translation early in my journey.

I'd also like to thank my practice partners in ARMA Provo/True Edge Academy who worked through the text with me, and especially Mark Driggs, Kristina Evans Feil, Rob Hyatt, Seth Pease, and Ben Smith, as well as the study group leaders who encouraged this weird line of investigation, Eli Combs and Stew Feil. Charles Deily of Schola Saint George/Boston Armizare likewise helped with some tricky questions of interpretation after I moved to Massachusetts.

This book would not have been possible without the support of the Morgan Library and Museum and Roger S. Wieck, the Melvin R. Seiden Curator and Department Head of Medieval and Renaissance Manuscripts. They accommodated my requests to visit the manuscript in person not once but twice: in October 2012 and again in September 2020. They also supported my plan to publish this book and provided much better scans than I could have gotten otherwise.

Ken Dietiker and Michael-Forest Meservy offered me valuable suggestions and criticism on the early Wiktenauer draft of this translation. For this publication, Carrie Patrick's editorial pen was greatly appreciated, as was Guy Windsor's detailed review. My beta readers were Jeffrey Cierniak and Jason James.

Finally, I'd like to thank Kendra Brown, who encouraged me to try working on my book projects full time and is thus directly responsible for this book's existence. She has also been very patient as it took over my life for several months.

Preface

In the spring of 2006, I broke my arm.

I was learning parkour at the time, and a fall of only about ten feet resulted in a left-side radial dislocation and ulnar fracture. Five hours of surgery later, I had a cast and a new steel plate, and my summer, which I'd planned to spend in OSUT training at Fort Sill, was now wide open. As I lay recovering in a hospital bed, I decided that it was time to study Fiore.

This didn't come out of the blue: I'd advanced to the rank of Free Scholar in ARMA the previous year, and the custom of our study group was that ranked members would explore a source that wasn't part of the curriculum and bring their findings back to teach to the group and enrich everyone's study. At this time, our group was very German-centric, and all I knew about Fiore was that it was the premier Italian source and that a lot of ARMA members liked to make jokes about it (and I didn't know anyone outside of ARMA yet). So, I decided I would be the guy who brought Fiore into the ARMA fold.

There were no books yet, so I went online and grabbed every Fiore resource I could locate. At the time, the only complete translation of one of his manuscripts was a translation of the Novati facsimile by Hermes Michelini of the Knights of the Wild Rose; Novati also provided the only available illustrations. Only fragments of the others were online: Matt Easton and Eleonora Litta of Schola Gladiatoria had released the Morgan preface and a few sections of the Getty (the preface, sword in one hand, sword in two hands, ax, and spear), and Rob Lovett and Mark Lancaster of the Exiles had released a few others from all three known manuscripts (the preface, wrestling, dagger, and dagger vs sword).

With only one set of illustrations, I decided out of the gate that the only way to study these sources was to put all the translations, as well as the transcriptions by Marco Rubboli and Luca Cesari, side-by-side on the same page. So, during my convalescence I carefully arranged them all play by play and put the resulting document

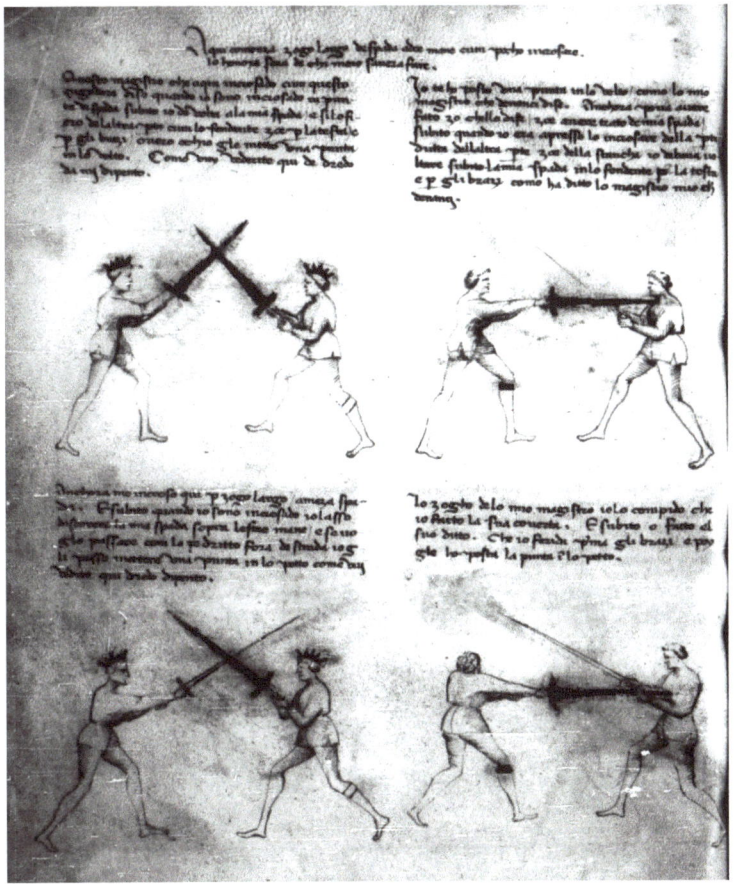

Fig. 1: Morgan 13ᵛ (scanned from microfilm)

1

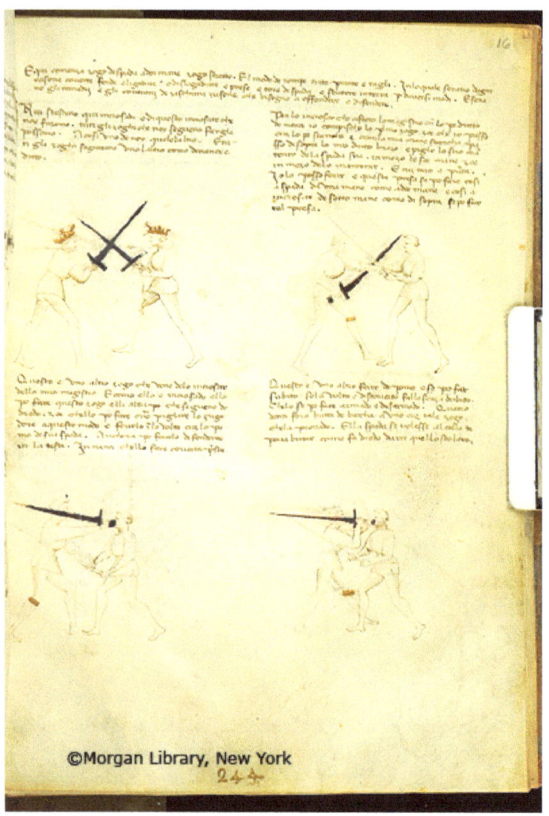

Fig. 2: Morgan 16ʳ (Christian Index of Christian Art)

in a big binder. The sword in one hand was of the most immediate interest to my one-armed self, and since ROB HYATT was working on his book on Henry de Sainct Didier at the time, we spent a lot of time that summer fencing longsword vs side sword.

Practicing on the side with various group members, I eventually worked through most of the text. I initially thought that my fluency in Spanish would suffice for actually reading Fiore, but I was sorely mistaken on that score and eventually took two years of Italian courses to get a better handle on the language. In the meantime, I blagged friends of mine, including BRIAN HUNT and PAOLO AMASSARI, into translating additional sections that weren't available online yet.

I also bought and scanned reels of microfilm with scratchy black and white scans of the other two manuscripts (fig. 1) to expand and revise my giant Fiore compilation. Then in 2008, I discovered that the Morgan manuscript had been scanned (in low resolution) as part of a partnership between the Morgan Library and Princeton University, and the scans were in a database called the Princeton Index of Christian Art. I was fortunate to still be a student, so I had easy access to this database and was able to see the Morgan in color for the first time (fig. 2). These were my first encounters with the Morgan, though they wouldn't be the last.

While the translation that you're holding in your hands might be said to have started all the way back in a hospital room in 2006, I didn't actually begin writing it until 2011. By this time, there were many more resources available for the Getty—a full translation by TOM LEONI, more sections from the Exiles, and even color scans published by MASSIMO MALIPIERO—but the Morgan had been largely ignored as an incomplete copy of the Getty (even after I had spread the Princeton scans as far and wide as I could).

I aimed to change that, and hatched a plan to publish not only the Morgan, but also a new translation of NOVATI. I envisioned this as a companion to LEONI's Getty translation, rounding out the Italian versions of Fiore's teachings. I worked on this during downtime at work for several months, but no sooner were the translations completed than LEONI released a new edition of his book which included (what was advertised as) a complete Morgan translation as a digital extra. This killed all my motivation at the time, and so I released the translation for free on Wiktenauer and considered the project over.

My fascination with Fiore was not over, though, and in 2012 I had the opportunity to visit the Morgan Library and Museum's reading room for the first time and see the manuscript in person. This was the first time I ever handled a medieval manuscript, and it cemented my affection for the shortest of Fiore's works. I would come back to my Fiore work many times over the subsequent years until in 2020, I realized that there was still no edition of this manuscript in print and that I could address that omission.

Fig 3: Studying the manuscript

I reached out to the Morgan, and they were amenable to such a publication but closed due to COVID-19. They finally reopened in September of that year, and I returned to the Sherman Fairchild Reading Room to take a second look at the manuscript. This visit was enormously helpful, and I made a number of observations that have never been described in the literature before. The Morgan also sent me quite enormous (850 dpi) scans to publish in the book, from which I was able to author a new transcription to replace the one done twenty years ago from microfilm. This transcription (along with nine further years of experience translating and editing the translations of others) allowed me to correct and improve my translation.

Despite my hospital resolution in 2006, I never became a real Fiorist. By 2007, I was taking on more teaching responsibilities and spending most of my time with the German sources that our curriculum was (hypothetically) based on, and in 2009 we officially broke away from ARMA and founded True Edge Academy. My work compiling Fiore led me to create a similar Liechtenauer concordance, and these two documents later formed the first foundations of Wiktenauer.

While I still read from the book of Fiore and carry it close to my heart to this day, the siren song of Liechtenauer's *Zettel* always calls me back. But the Fiore community is important to me, and I hope this book will help us take another step forward.

Michael Chidester
4 April 2021

THE FLOWER OF BATTLE: MS M 383

Introduction

Fiore Furlano de'i Liberi de Cividale d'Austria (fl. 1381 - 1409) was a late 14th century knight, diplomat, and fencing master. He was born in Cividale del Friuli, a town in the Patriarchal State of Aquileia (in the Friuli region of modern-day Italy), the son of Benedetto and scion of a Liberi house of Premariacco.[1] The term *Liberi*, while potentially merely a surname, more probably indicates that his family had Imperial immediacy (*Reichsunmittelbarkeit*), either as part of the *nobili liberi* (*Edelfrei*, "free nobles"), the Germanic unindentured knightly class which formed the lower tier of nobility in the Middle Ages, or possibly of the rising class of Imperial Free Knights.[2, 3, 4] It has been suggested by various historians that Fiore and Benedetto were descended from Cristallo dei Liberi of Premariacco, who was granted immediacy in 1110 by Holy Roman Emperor Heinrich V,[5, 6, 7] but this has yet to be proven.[8]

Fiore wrote that he had a natural inclination to the martial arts and began training at a young age, ultimately studying with "countless" masters from both Italic and Germanic lands.[9] He had ample opportunity to interact with both, being born in the Holy Roman Empire and later traveling widely in the northern Italian states. Unfortunately, not all of these encounters were friendly: Fiore wrote of meeting many "false" or unworthy masters in his travels, most of whom lacked even the limited skill he'd expect in a good student.[10] He further mentions that on five separate occasions he was forced to fight duels for his honor against certain of these masters who he described as envious because he refused to teach them his art; the duels were

---

[1] Morgan, folio 1ʳ (pp 40-41). Whenever page numbers are given for Morgan citations, they refer to the transcription and translation chapter of this book. Many of these footnotes could equally cite the Getty or Pisani Dossi text, but I've chosen to limit them to the Morgan both since it is the primary subject of this book and since the text can be readily consulted.

[2] He is never given such a surname in any contemporary records of his life, and the term only appears when introducing his family in his own treatises.

[3] MONDSCHEIN 2011B 11.

[4] RUSS HOWE. "Fiore dei Liberi: Origins and Motivations". *Journal of Western Martial Art*. Electronic Journals of Martial Arts and Sciences, 2008. https://ejmas.com/jwma/articles/2008/jwmaart_howe_0808.htm. Retrieved 7 April 2021.

[5] GIUSTO FONTANINI. *Della Eloquenza italiana di monsignor Giusto Fontanini*, vol. 3. R. Bernabò, 1736. pp 274-276.

[6] GIAN GUISEPPE LIRUTI. *Notizie delle vite ed opere scritte da' letterati del Friuli*, vol. 4. Alvisopoli, 1830. p 27.

[7] NOVATI 15-16.

[8] MALIPIERO 80.

[9] Morgan, folio 1ʳ (pp 40-41).

[10] NOVATI 15-16.

all fought with sharp swords, unarmored except for gambesons and chamois gloves, and he won each without injury.[11]

Writing very little about his own career as a commander and master at arms, Fiore laid out his credentials for his readers in other ways. He stated that foremost among the masters who trained him was one Johane dicto Suueno, who he notes was a disciple of Nicholai de Toblem;[12] unfortunately, both names are given in Latin so there is little we can conclude about them other than that they were probably among the Italians and Germans he alludes to, and that one or both were well known in Fiore's time. He further offered an extensive list of the famous *condottieri* that he trained, including Piero Paolo del Verde (Peter von Grünen), Niccolo Unricilino, Galeazzo Cattaneo dei Grumelli (Galeazzo Gonzaga da Mantova), Lancillotto Beccaria di Pavia, Giovannino da Baggio di Milano, and Azzone di Castelbarco, and also highlights some of their martial exploits.[13]

The only known historical mentions of Fiore appear in connection with the Aquileian War of Succession, which erupted in 1381 as a coalition of secular nobles from Udine and surrounding towns sought to remove the newly appointed Patriarch (prince-bishop of Aquileia), Cardinal Philippe II d'Alençon. Fiore seems to have supported the secular nobility against the Cardinal; he traveled to Udine in 1383 and was granted residency in the city on 3 August.[14] On 30 September, the high council tasked him with inspection and maintenance of the city's weapons, including the artillery pieces defending Udine (large crossbows and catapults).[15] In February of 1384, he was assigned the task of recruiting a mercenary company to augment Udine's forces and leading them back to the city.[16] This task seems to have been accomplished in three months or less, as on 23 May he appeared before the high council again and was sworn in as a sort of magistrate charged with keeping the peace in one of the city's districts. After May 1384, the historical record is silent on Fiore's activities; the war continued until a new Patriarch was appointed in 1389 and a peace settlement was reached, but it's unclear if Fiore remained involved for the duration. Given that he appears in council records four times in 1383-84, it would be quite odd for him to be completely unmentioned over the subsequent five

---

[11] Morgan, folio 1ᵛ (pp 42-43).
[12] NOVATI, carta 1ᴬ.
[13] Morgan, folio 1ʳᵛ (pp 40-43).
[14] MALIPIERO 84.
[15] MALIPIERO 85.
[16] MALIPIERO 86.

years if he remained,[17] and since his absence from records coincides with a proclamation in July of that year demanding that Udine cease hostilities or face harsh repercussions, it seems more likely that he moved on.

Based on his autobiographical account, Fiore traveled a good deal in northern Italy, teaching fencing and training men for duels. He seems to have been in Perugia in 1381 in this capacity, when his student Peter von Grünen likely fought a duel with Peter Kornwald.[18]

In 1395, he can be placed in Padua training the mercenary captain Galeazzo Gonzaga of Mantua for a duel with the French marshal Jean II le Maingre (who went by the war name "Boucicaut").[19] Galeazzo made the challenge when Boucicaut called into question the valor of Italians at the royal court of France, and the duel was ultimately set for Padua on 22 August. Both Francesco Novello da Carrara, Lord of Padua, and Francesco Gonzaga, Lord of Mantua, were in attendance. The duel was to begin with spears on horseback, but Boucicaut became impatient and dismounted, attacking Galeazzo before he could mount his own horse. Galeazzo landed a solid blow on the Frenchman's helmet but was subsequently disarmed. At this point, Boucicaut called for his poleaxe but the lords intervened to end the duel.[20]

Fiore surfaces again in Pavia in 1399, this time training Giovannino da Baggio for a duel with a German squire named Sirano. It was fought on 24 June and attended by Gian Galeazzo Visconti, Duke of Milan, as well as the Duchess and other nobles. The duel was to consist of three bouts of mounted lance followed by three bouts each of dismounted poleaxe, estoc, and dagger. They ultimately rode two additional passes and on the fifth, Baggio impaled Sirano's horse through the chest, slaying the horse but losing his lance in the process. They fought the other nine bouts as scheduled, and due to the strength of their armor, both combatants reportedly emerged from these exchanges uninjured.[21, 22]

Fiore was likely involved in at least one other duel that year, that of his final student Azzone di Castelbarco and Giovanni degli Ordelaffi, as the latter is known to have died in 1399.[23]

After Castelbarco's duel, Fiore's activities are unclear. Based on the allegiances of the nobles that he trained in the 1390s, he seems to

---

[17] MALIPIERO 85-88.
[18] This is the only point when both men are known to have been in Perugia at the same time; Verde died soon after this in 1385.
[19] See pages 14-19.
[20] MALIPIERO 55-58.
[21] MALIPIERO 84.
[22] See pages 19-22.
[23] MALIPIERO 97.

have been associated with the ducal court of Milan in the latter part of his career. Sometime in the first years of the 1400s, Fiore composed a fencing treatise in Italian and Latin called "The Flower of Battle" (rendered variously as *Fior di Battaglia*, *Florius de Arte Luctandi*, and *Flos Duellatorum*). The briefest version of the text is dated to 1409 and indicates that it was a labor of six months and great personal effort;[24] as evidence suggests that at least two longer versions were composed some time before this,[25] we may assume that he devoted a considerable amount of time to writing during this decade.

Beyond this, nothing certain is known of Fiore's activities in the 15th century. FRANCESCO NOVATI and LUIGI ZANUTTO both assume that some time before 1409 he accepted an appointment as court fencing master to Niccolò III d'Este, Marquis of Ferrara, Modena, and Parma; presumably he would have made this change when Milan fell into disarray in 1402, though ZANUTTO went so far as to speculate that he trained Niccolò for his 1399 passage at arms.[26] The records of the d'Este library indicate the presence of two versions of "the Flower of Battle", but it seems more likely that the manuscripts were written as a diplomatic gift to Ferrara from Milan when they made peace in 1404.[27]

C. A. BLENGINI DI TORRICELLA stated that late in life, Fiore made his way to Paris, where he could be placed teaching fencing in 1418 and creating a copy of a fencing manual located there in 1420. Though BLENGINI attributes these facts to NOVATI, no publication verifying them has yet been located and this anecdote may be entirely spurious.[28]

The time and place of Fiore's death remain unknown.

Despite the extent and complexity of his writings, Fiore de'i Liberi does not seem to have been a very significant master in the evolution of fencing in Central Europe. Even so, there are a number of later treatises which bear strong resemblance to his work,

---

[24] NOVATI, carta 1^B

[25] Fiore states in the preface to the Pisani Dossi that he had studied combat for fifty years, whereas the comparable statement in the Morgan and the Getty mention the slightly shorter "forty years and more". See Morgan 2^r (pp 44-45).

[26] ZANUTTO 211-212.

[27] MONDSCHEIN 2011B 12.

[28] In 1907, fencing master C. A. BLENGINI DI TORRICELLA wrote that "In 1904, a historical work by Francesco Novati, Director of the Academy in Milano and Gaffuri, Director of the graphical institute in Bergamo was published... These two prominent scholars uncovered documents, found in different archives, ...*Rules for Fencing* were printed by Fiore dei Liberi in 1420... And how could then dei Liberi have taught fencing lessons in Paris in 1418?" (translated from Norwegian by Roger Norling). See BLENGINI, DI TORRICELLA C. A. *Haandbog i Fægtning med Floret, Kaarde, Sabel, Forsvar med Sabel mod Bajonet og Sabelhugning tilhest: Med forklarende Tegninger og en Oversigt over Fægtekunstens Historie og Udvikling*. 1907. p 28.

including the writings of Philippo di Vadi and Ludwig VI von Eyb. This may be due to the direct influence of Fiore or his writings, or it may instead indicate that the older tradition of Johane and Nicholai survived and spread outside of Fiore's direct line.

The Flower of Battle

The d'Este family owned three manuscripts by Fiore during the 15th century,[29] and a total of four copies survive to the present. Of these, the Getty and the Pisani Dossi are both dedicated to Niccolò III d'Este and state that they were written at his order. The Morgan, which is the subject of this book, lacks a dedication and states that it was laid out according to Fiore's own intelligence, while the Paris lost any dedication it might have had along with its preface. Each of the extant copies of the *Flower of Battle* is unique, though each of these pairs contains strong similarities to each other in order of presentation.

In addition, Philippo di Vadi's manuscript from the 1480s (Rome), whose second half is essentially a redaction of the *Flower of Battle*, provides a valuable fifth point of reference when considering Fiore's teachings.

The major sections of the work include: *abrazare* or grappling; *daga*, including both unarmed defenses against the dagger and plays of dagger against dagger; *spada a un mano*, the use of the sword in one hand (also called "the sword without the buckler"); *spada a dui mani*, the use of the sword in two hands; *spada en arme*, the use of the sword in armor; *azza*, plays of the poleaxe in armor; *lancia*, spear and staff plays; and mounted combat (including the spear, the sword, and mounted grappling). Brief bridging sections serve to connect each of these, covering such topics as sword vs. dagger and the use of the javelin against a man on horseback.

---

[29] There are two records in the 1436 and 1467 catalogs which mention Fiore, but only one of the manuscript descriptions is similar between the catalogs. The 1436 catalog lists one unbound Latin manuscript and one Italian manuscript bound in red leather; the 1467 catalog lists two Latin manuscripts, one of which was 15 unbound folia (probably the same as the one from 1436) and one of which was 58 folia bound in white leather. From this, we might speculate that the Getty was present in 1436, the Paris in 1467, and the third (very short) manuscript is currently unknown to us. If there were an error in the 1467 catalog (or its transcription), then the unknown manuscript could be the Pisani Dossi, which currently consists of 35 unbound folia and which begins with a Latin preface before switching to Italian.

The Flower of Battle: MS M 383

Dueling in the 15ᵗʰ Century

One of the first things you might notice about the manuscript is that it does not follow the sequence of teachings you would expect based on the Getty and Pisani Dossi. Those two manuscripts structure their plays in a way that seems pedagogical, beginning with unarmored wrestling, dagger, and sword before moving on to armored sword, ax, and spear, and finally mounted spear, sword, and wrestling.

Our manuscript, on the other hand, reverses that order. It begins with the plays of fencing and wrestling on horseback and then moves to the spear and then the sword. (According to the plan given in the preface, wrestling and dagger were to come last of all.)

So does this represent an alternate pedagogy? After all, horsemanship is even more of a foundational skill for a young warrior than wrestling is, and the play of the spear is much simpler than the plays of the sword (and, in fact, the spear is taught before the sword in the Pisani Dossi).

It might, but this order of teachings is actually much more common across the corpus of fencing treatises than is that of the Getty and the Pisani Dossi. It is very apparent in the Vienna (1420s), colloquially referred to as *Die Blume des Kampfes*, but that text is closely related to the works of Fiore de'i Liberi so it comes as no surprise. This sequence is also used in the many manuscripts based on the 1460s treatise of German master Paulus Kal,[30] and the sequence of horse, spear, sword, and dagger is also the basic framework of the armored fencing of the German grandmaster Johannes Liechtenauer himself.[31] Other treatises, such as

Fig. 1: The end of a judicial duel
Ms. Français 2258, 23ᵛ (ca. 1450)
Bibliothèque nationale de France

---

[30] Paulus Kal's work is preserved in at least five substantial copies: Bayerische Staatsbibliothek, Cgm 1507 (1460s); Biblioteca Universitaria di Bologna, ms. 1825 (1460s); Universitäts- und Forschungsbibliothek Erfurt/Gotha, ms. Chart. B. 1021 (1473-1503); Kunsthistorisches Museum, KK5126 (ca. 1480); Zentralbibliothek Solothurn, Cod. S. 554 (1506-14). At least three other small fragments have surfaced at auction in the past century, suggesting the previous existence other works that have been dismembered. For a translation of Cgm 1507, see CHRISTIAN HENRY TOBLER. *In Service of the Duke: The 15th Century Fighting Treatise of Paulus Kal*. Highland Village, TX: Chivalry Bookshelf, 2006. See also "Paulus Kal". *Wiktenauer*. https://wiktenauer.com/wiki/Paulus_Kal.

[31] The unmodified *Zettel* of Johannes Liechtenauer exists in at least sixteen substantial copies, and it is also quoted by various later fencing masters. For further discussion, see HANS-PETER HILS. *Meister Johann Liechtenauers Kunst des langen Schwertes*. P. Lang, 1985. See also "Johannes Liechtenauer". *Wiktenauer*. https://wiktenauer.com/wiki/Johannes_Liechtenauer.

Fig. 2: A trial by combat following the Frankish rite
Ms. Chart A. 558, 42ʳ (ca. 1448)
Forschungsbibliothek Gotha

the *Gladiatoria* group[32] of the 1430s and the 1480s treatises of Fiore's literary successor Phillipo di Vadi,[33] follow a truncated version of it.

The reason this pattern is so ubiquitous is because it has a very real, concrete meaning: this is the sequence of actions common in the martial dueling of the 14th and 15th centuries. A thorough discussion of martial dueling would require an entire book to even begin to cover, but I will attempt the briefest of overviews here.[34]

These duels were conceived of as battles; the word itself comes the *duellum*, which was a word in Old Latin that retained poetic use after the pronunciation shifted to *bellum* ("war") in Classical Latin. This word was revived in the Late Medieval period because it was misinterpreted as something like *duo bellum*, a war between two. The word *duello* or *duel* had little currency even into the 15th century, and terms like "single combat" and "battle of two" were far more common.

This practice was entirely divorced from the Saxon legal tradition in which certain types of cases could be resolved by battles in leather costumes with shields and clubs or swords.

Instead, this arose in the late 13th century from ways of settling disputes in camp between knights or men at arms on campaign. The army would build an enclosure and the disputants would settle

---

[32] This is a complex of four manuscripts: Yale Center for British Art, ms. U860.F46 1450 (1430s); Kunsthistorisches Museum, ms. KK5013; Biblioteka Jagiellońska, ms. germ. quart. 16 (1440s); Herzog August Bibliothek, cod. Guelf. 78. 2. Aug. 2° (1465-80). A substantial fragment of one section is also found in Musée national du Moyen Âge, Cl. 23842 (ca. 1490), and the group may be based on the illustrations of Universitätsbibliothek Augsburg, Cod.I.6.4°.2 (1420s). See DIERK HAGEDORN and BARTŁOMIEJ WALCZAK. *Gladiatoria. New Haven - MS U860.F46 1450*. Herne: VS-Books, 2015. See also "Gladiatoria group". *Wiktenauer*. https://wiktenauer.com/wiki/Gladiatoria_group.

[33] Vadi's works are primarily known from Rome (1482-5), but there is evidence of one or even two other manuscripts which have yet to be properly documented. For a translation of the known copy, see GUY WINDSOR. *The Art of Sword Fighting in Earnest*. Spada Press, 2018. See also "Philippo di Vadi". *Wiktenauer*. https://wiktenauer.com/wiki/Philippo_di_Vadi.

[34] Discussion of dueling under the *jus belli* or Law of War has been sadly neglected in the literature, and it is frequently conflated with the tradition of trial by combat, muddying the waters for both. The principal work on the subject is MUHLBERGER, STEVEN. *Deeds of Arms*. Highland Village, TX: Chivalry Bookshelf, 2005; this book is concerned only with the Anglo-French sphere, though. For Central Europe, there is brief discussion in ARIELLA ELEMA. "Tradition, Innovation, Re-enactment: Hans Talhoffer's Unusual Weapons". *Acta Periodica Duellatorum*. 7(1): 3-25, 2019. doi:10.2478/apd-2019-0001 and MARCO CAVINA. "Science of Duel and Science of Honour in the Modern Age: The Construction of a New Science between Customs, Jurisprudence, Literature and Philosophy". *Late Medieval and Early Modern Fight Books*: 571-593. Brill, 2016. doi: 10.1163/9789004324725_020, and at only slightly more in STEVEN HUGHES. "Soldiers and Gentlemen: The Rise of the Duel in Renaissance Italy". *Journal of Medieval Military History* V: 99-152. Boydell & Brewer, 2007. I am currently preparing a book on this subject.

their differences in front of their fellow fighters, under the supervision of their captains. This practice was adopted into civil law as veterans of the Hundred Years War returned from campaigns and wanted to settle their grievances the way they were accustomed to.

These duels were conceived of as battles, and they defy a lot of expectations set by customs of later centuries. The objective was to *overcome and defeat* your opponent, not necessarily to kill him. Indeed, it often happened that one duelist would be taken prisoner by the other in the course of the duel and then ransomed back to his people, just like on a real battlefield. Other times, the duels were arranged in similar manner to exhibitions, with clearly defined beginnings and ends (such as breaking a certain number of lances or fighting a certain number of passes). Indeed, many fights that are described by chroniclers and historians as "feats" or "deeds of arms" are equally described by others as duels.

Often duelists did die, though, and the custom was designed to allow for this outcome. The challenged party would be required to locate both a prince willing to sponsor the field and a judge willing to supervise the fighting. This gave the duel official legal standing and established that if one of the duelists killed the other, it was not a criminal act but rather part of the legal proceeding.

Duels in which concessions were made for safety—though they would still be dangerous and even lethal from time to time—were often characterized as *ad piacenza* ("to satisfaction"), whereas duels in which no concessions to safety were made—with death a likely, though still not required, outcome—were *ad oltranza* ("to the extreme").

Fiore helpfully offers some examples of this in the form of boasting about his famous students:[35]

> *"And for some who have been my students and have been obliged to combat in the list, I want to name them and make a remembrance here...*
>
> *"The notable, valiant, and gallant knight Sir Galeazzo Cattaneo dei Grumelli, called da Mantua, who was obliged to combat with the valiant knight Sir Boucicaut of France, and the field was at Padua.*
>
> *"Also the valiant squire Lancillotto Beccaria of Pavia, who made six thrusts of sharpened iron lance on horseback against the valiant knight Sir Balthasar (a German). They were also obliged to combat in the list, and this was at Imola.*

---

[35] Morgan 1ʳᵛ (pp 40-43).

> *"Also the valiant squire Giovannino da Baggio of Milan, who made three thrusts of sharpened iron lance on horseback against the valiant squire Sirano (a German) in the castle in Pavia. And then, dismounted, he made three blows of axe, and three blows of sword, and three blows of dagger, in the presence of the most noble lord Duke of Milan, and of the lady Duchess, and of countless other lords and ladies."*

No other description of the duel between Lancillotto Beccaria and Balthasar von Braunschweig-Grubenhagen has yet surfaced, but fortunately, we have detailed accounts of the other two from people who were in attendance. Even more fortunately, between the two, they capture norms for the conduct of duels very well.

Galeazzo Cattaneo dei Grumelli

This description of the duel between Galeazzo da Mantua and Boucicaut (Jean II le Maingre), which occurred on 22 August 1395, was written by Galeazzo and Andrea Gatteri in their chronicle of the Carrara lords of Padua. It offers an example of the procedural nature of a duel outside of the fighting itself: the challenge, the difficulty in securing a field and a judge to make the duel legal, the ritual surrounding the duel itself, and so on:[36]

> *"The valiant knight Sir Galeazzo da Mantua, finding himself in France in the king's court where there were a multitude of barons (as it sometimes is found to be in royal courts) and speaking here of one thing and here of another, and here of Italians and here of Germans, it seems that Sir Boucicaut said certain words against Italians, and Sir Galeazzo caught those words and said that he spoke poorly, and they quarreled with each other in words, and many barons intervened to pacify them.*
>
> *"Ultimately nothing could be done because, each one wanting to keep his honor and show that what he had said was true, Sir Galeazzo asked for battle, and Sir Boucicaut asked him to come and find a field since the king did not want to grant it to him, and so he agreed that he would come and find a field...*
>
> *"Sir Galeazzo left and came to Mantua and asked the lord for a field, but he didn't want to grant one, so he came to Padua and the lord voluntarily had and granted the field, telling Sir*

---

[36] Translation is my own, based on the transcription in MALIPIERO 55-58.

Boucicaut to be in Padua on 15 August if he wanted to maintain in person what he had said, and he came to Padua...

"To notify you readers, I say that the fame of these two notable knights who had to fight was everywhere, so that many gentlemen and lords came to see such a thing, and hearing the great and beautiful preparations that the lord of Padua made, everyone tried to come and see: the lord of Mantua, the lord Carlo Malatesta, the lord of Faenza, the lord of Forlì, Sir Tristan of Savorgnan, Certesin of Zuan Lion of Cremona, captain of Vicenza, and many gentlemen from Venetia and Treviso. Certainly there were over twelve thousand people that came to see.

"But to notify you about the complete preparations made by the magnificent Lord of Padua in the middle of his square, I say that the apparatus was made like this: first, a square list was built from the middle of the gate, across the court, and at the end of the list (next to the banquet hall) he built a large chair covered with a cloth with Boucicaut's heraldry; behind this, like the deck of a galley, stood a balustrade as long as the list, on which his family stood. In the middle of the road which goes from the cathedral, a wooden tower was built, on which the lord placed his crossbowmen. Under the balcony of the court was built a terrace, hanging from the gate of the court to the tower, and there the lords who had come with the gentlemen from Venice were, and on the other side of the court, outside, above, and beside the house of the Via de Strà, another terrace was built where those gentlemen from Bologna, Ferrara, Triviso, and Friuli stood.

"Around the list on three sides were made bleachers that were 3 1/2 feet tall, all of boards and closed on the back, on which were all the citizens as well as armed crossbowmen to guard the combatants. Across the square, from one side to the other, was made a hanging terrace on which stood people of all conditions; terraces were also made on the roofs of the houses, which were all full of people, and seats on this balcony were rented at three ducats. Perhaps never have we had so many people to see such a thing. From the well of the square to the court, a passageway was left that could be used to ride to

Fig. 3: Preparations for a duel
Ms. Français 2258, 9ʳ (ca. 1450)
Bibliothèque nationale de France

The Flower of Battle: MS M 383

Fig. 4: *Swearing the oath*
Ms. Français 2258, 16ʳ (ca. 1450)
Bibliothèque nationale de France

the fence. Above the court's gate was hung a hawk, under which stood the lord of Padua and that of Mantua. Along the fence toward the court was made another passageway in which were the lord's mercenaries, armed but without spears. Thus were the people settled, watching for the baron combatants to come.

"Sir Boucicaut came into the field on Sunday, 22 August 1395. While the crowd was attentive without any noise, Sir Boucicaut's company began to come, without any instruments, by the way of da Lion. He came in this way: three armored coursers were led one behind the other, the first all armored with mail that looked like silver, the second armored with a multi-colored armament made of snake scales, the third with a velvet blanket embroidered with his heraldry, that is, a great crane, half red and half white; after this came two pages on coursers with lances in hand, one with a helmet on his head and the other a small shield on his back; behind them came Sir Boucicaut on a black courser, with his body so armored that he looked like Saint George.

"In such order they came up to the door of the fence, where Sir Michiel da Rabara, Sir Morando da Porcilia, and Polo da Lion (along with other gentlemen) presented him a missal, and he swore a sacramental oath according to the custom of combatants, and also that he would do no more than what the lord of Padua and the lord of Mantua wanted, who were present there and who accompanied him from the coast. Thus he entered the fence, dismounted from his horse, and went to his chair. He sent away his display horses and all his horses except one, and so he waited for his opponent, Galeazzo da Mantua, who was located in the house of da Lion.

"He too was ready and came along Via de Strà with three horses covered with silk, though not as magnificently as Sir Boucicaut, and he too was accompanied up to the fence by the previously-mentioned gentlemen, and there he was made to swear by them as above. Having done this, he entered the fence and went towards his pavilion, which was in front of that of Sir Boucicaut, and when he reached the middle, he bowed and honored Sir Boucicaut, and was much commended by everyone. Then he dismounted, sent his horses away (retaining one), and entered the pavilion and closed himself inside.

"While this was the case, the lord of Padua sent the lord of Mantua to try to make them make peace together, but he could do nothing.

"Those who were seeking on behalf of Sir Boucicaut were the gentlemen written above: Sir Calzin Turmello and Sir Francesco Buzzaccarin and three French knights; on behalf of Sir Galeazzo, Sir Michiel di Rabara, Sir Morando da Porzia, and Sir Polo da Lion. They went from one knight to the other but could do nothing, and instead spent hours measuring spears and axes and other armaments.

Fig. 5: Mounted fighter
Ms. Ludwig XV 13, 41r (ca. 1400)
J. Paul Getty Museum

"Seeing that there would be no agreement, Sir Boucicaut did not want to delay any further, so he had his courser come and mounted it, with the helmet on his head, the shield in front of his chest, and a lance in his hand. He began to cross the field, waiting for Sir Galeazzo to mount his horse. Sir Galeazzo took his horse by the reins and waited for the glove of battle to be thrown, but when Sir Boucicaut saw that Sir Galeazzo did not mount his horse, he dismounted from his horse and went to his chair and had his spurs removed (it was in the agreement between them that if one went to his seat, his companion would not trouble him).

Fig. 6: A thrust with a spear
Ms. Ludwig XV 13, 40r (ca. 1400)
J. Paul Getty Museum

"Spurs removed, Sir Boucicaut took the lance in his hand and boldly went to meet Sir Galeazzo, believing that the glove had been thrown already. Sir Michiel di Rabara, seeing Sir Galeazzo and Sir Boucicaut come towards each other on foot with lance in hand, quickly threw the glove of battle.

"Sir Galeazzo had the visor of his helmet raised and, seeing Boucicaut come in like manner with his lance in his hand, he struck the camail of Sir Boucicaut's helmet in such a way that he knocked him back three steps. Sir Boucicaut threw

Fig. 7: Stance with the ax
Ms. Ludwig XV 13, 35ᵛ (ca. 1400)
J. Paul Getty Museum

away his lance, put his hands on the lance of Sir Galeazzo, and broke it. Then he put hand to his ax that he had at his side, and he took it up with both hands and went toward Sir Galeazzo to give it to him good.

"But the lord of Mantua with some of the lord's mercenaries ran towards him and grabbed him from the sides, saying, "No more, you have done much! Don't you want to attend what you promised and swore on the missal, to do whatever the lord of Padua and I wish?" On the other side, the lord of Padua and his mercenaries had gone to Sir Galeazzo in the same way and taken him in the same way, and said those same words that were said to Sir Boucicaut. After many words, they finally pacified them together and led them to the middle of the field, and they took off their helmets and embraced each other and kissed each other with great joy."

Here we have a basic template for a duel. One knight offended the other and insults were exchanged, leading to a formal challenge. The wording seems to suggest that Galeazzo issued the challenge, since Boucicaut first sought a field in France but was unable to find one. If the defendant fails to find a field, the challenger may seek one (though the defendant has the right to reject it—the defendant controls all the conditions of a duel, since presumably he didn't want to fight the duel in the first place), and Galeazzo secured a field in Italy.

Duels were explicitly public proceedings, so the dueling field or list was built in the public square; a duel between famous knights like this drew a massive crowd, which the prince hosting the duel also had to accommodate. On the day of the duel, the fighters assembled at the field, took oaths, and made themselves ready in their own tents. The last-minute negotiations to avoid the duel were also typical in these situations.

The duel was to begin on horseback, but Boucicaut's impatience resulted in beginning the duel on foot. The duel ended as quickly as it began: the prince hosting the field was not just an observer, but the ultimate authority in the proceeding, and he retained the right to end a duel at any time. In this case, he allowed each fighter to make one impressive martial feat, and then declared an end to the fight. From this description, it's unclear if the knights were using sharp weapons or fighting "in the extreme", but that wouldn't matter if the prince decided the duel should end.

Giovannino da Baggio

The duel between Giovannino da Baggio and Sirano, which occurred on 24 June 1399, lacks the lengthy preamble we just saw, but continues through much more action:[37]

> *"Being intended that Sirano the German was to make three strikes with the sharp lance, three of ax, three of sword, and three of dagger, he personally selected Giovannino da Baggio of Milan as adversary, a youth of courage, good looks, and totally obscure lineage, because he was satisfied of his desire.*
>
> *"At the 19th hour of the 24th of the present June, a signal was called in the field in the castle of Pavia, at which they entered separately to the music of flutes (which seemed to echo from the sky) with a magnificent entourage, illustrious men of valor; among the knights who accompanied Giovannino were Count Corrado di Altimbegh, Count Ugho Sciran, Lord Galeazzo de Porris, and Lord Franchino de Ruschis.*
>
> *"They ran here and there on the field, in order to accustom the horses to run through and move with assurance. Then the most illustrious Lady Duchess and the illustrious Lord, and the father of your magnificence, Count Antonio di Urbino, sat to watch in a grandstand arrayed in precious fabric and with golden supports. Together with them were others that I have left out because they aren't worth remembering... In the meantime, Mars inflamed the opposition and discord between the fighters so that they wouldn't desist from the emprise, but were very violently animated toward murder.*
>
> *"Sirano, whose horse was unarmored, asked that Giovannino's horse armor be stripped. After time lost in words, the 20th hour sounded and called them to the duel. They put on their helmets, took up their sharp lances, and their shining armor and shields didn't hide their standards.*
>
> *"After having deemed the ground safe, they ferociously spurred their horses with spurs, clashing at full speed but with no result.*

---

[37] Translation is my own from MALIPIERO 94-96; this is a modern Italian translation, but MALIPIERO also offers a transcription of the Latin on pp 503-505. He cites a collection of records and documents from late 14th century Urbino which was published by G. FRANCESCHINI, but doesn't indicate who the original author of the account was.

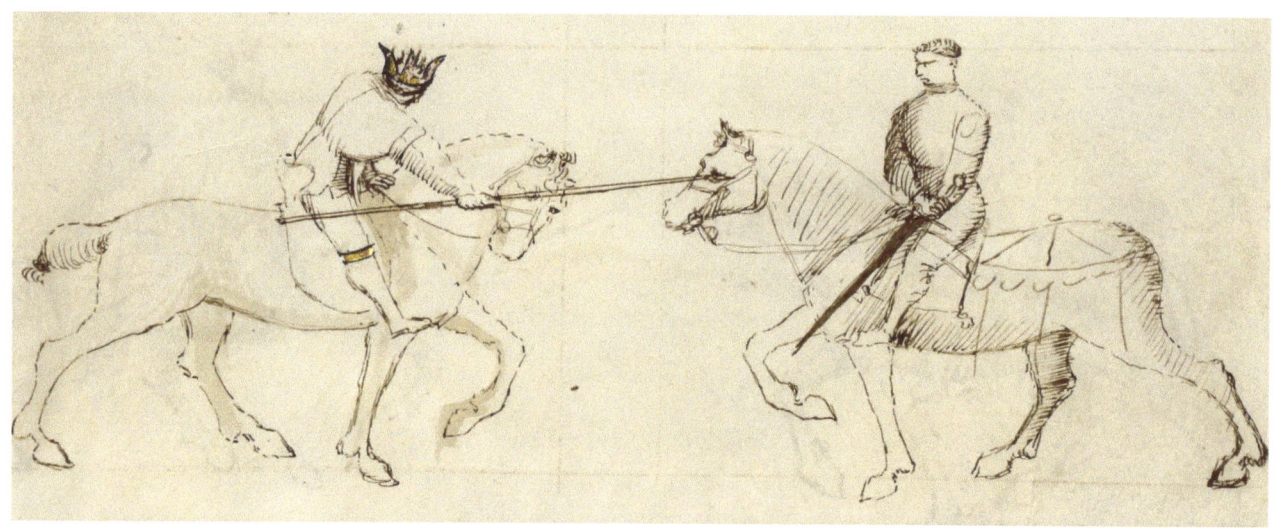

Fig. 8: Attacking the opponent's horse
Ms. Ludwig XV 13, 43ʳ (ca. 1400)
J. Paul Getty Museum

Fig. 9: An attack with the ax
Ms. Ludwig XV 13, 36ᵛ (ca. 1400)
J. Paul Getty Museum

In the second and third assaults, Giovannino struck Sirano in the shoulder, but without any wound. On the fourth course, Sirano was infuriated and hit Giovannino on the helmet so hard that it bowed toward his horse's leg.

"After these assaults, wanting by mutual agreement to clash in one final pass and driven by great ardor, they increased the distance to be covered and spurred their horses on with the sharp points of their spurs, and they hurled themselves bravely at free rein against each other. Sirano's horse took an oblique course, no doubt because of the violence, refusing to run straight as the two brave men clashed with mutual peril. Sirano's steed, I think more from desire for its own ruin than from fear, moved from the straight line of the course and received the iron in its flank, with the rod between its bowels and seen coming out over its buttocks. Thus, Sirano lay stretched out on the ground while Giovannino's spear remained stuck in the horse and couldn't be removed by the arm of this Giovannino, who fell to the ground from the force of his pulling. And thus, oh magnificent lord, here was the end of the clash on horses which occupied them for half an hour. Meanwhile, in the middle of the field, two silk-covered chairs were placed where the contenders sat and rested.

"Shortly afterward, they took up axes and each occupied the field, moving against each other: Giovannino hit Sirano on the head and Sirano hit him with the ax handle on the shoulder (he had, in fact, gotten too close to his enemy). They met again, moving with large steps through the space of forty feet, and Sirano struck Giovannino in the thigh and Giovannino struck Sirano in the head... But Giovannino, outraged by the blow to the thigh, struck Sirano with a similar blow in the shin from which Sirano was somewhat hobbled. Sirano in turn hit Giovannino in the head but, since there was no wound, after a short interval they took the swords.

"The shape of their swords was difference from modern ones, and in fact the blades had a sturdy and thin base and they played by striking only with the point. With great strength and commitment, they struck each other with both hands in the same way for three passes.

"After taking another interval of resting time, fighting with the dagger produced a great clangor. Through one blow and another of the same type of weapon, they left the aforementioned field, praising God that no trace of blood appeared on either of them.

"These, these, my magnificent lord, are the armors which protect human blood, which repel such powerful blows, which blunt such sharp points! I wish you had similar ones; they say they can hardly fear the lightning from Vulcan.

"In fact, these duelists wore the following protections: iron covering for the legs and a steel lorica, over which they placed a corslet tested against all blows and suitable to protect against any threat, whose iron chest covered down to the crotch, extending transversely with finger-sized lames."

This duel format, with a set number of passes with specific weapons, is fairly common. Oddly, it sounds like Sirano planned the duel first and then selected an opponent. This description gives us the full

Fig. 10: A parry with the sword
Ms. Ludwig XV 13, 33ʳ (ca. 1400)
J. Paul Getty Museum

Fig. 11: Defense with the dagger
Ms. Ludwig XV 13, 38ʳ (ca. 1400)
J. Paul Getty Museum

progression of mounted lance and dismounted ax, sword, and dagger. It also describes the equipment they used in some detail.

Going to the Woods

There is another series of duels mentioned in Fiore's preface, of course:[38]

Fig. 12: Unarmored fencing
Ms. Ludwig XV 13, 25ʳ (ca. 1400)
J. Paul Getty Museum

"And most of all, I have been wary of fencing masters and their students. And out of envy, they (that is, the masters) challenged me to play at swords of sharpened edge and point, in arming jacket but without any other armor save for a pair of chamois gloves. And all this just because I didn't want to practice with them, nor did I want to teach them anything of my art.

"And this incident, that I was so challenged, occurred five times. And five times I had to play for my honor in strange places, without relatives and without friends, having no hope in anything other than in God, in the art, and in myself, Fiore, and in my sword. And by the grace of God, I, Fiore, remained with honor and without wounds on my body."

What does this refer to? How does it fit into the dueling code mentioned above? The short answer is that it doesn't, and the slightly longer answer is that establishing the context of 14th-15th century unarmored fencing has proven extremely difficult and there are still no easy explanations.

However, Paride del Pozzo[39] (1410-91) cracks open the door slightly in his 1472 treatise on dueling laws and customs.[40] Pozzo was a jurist and legal professor, and in this work he sought to not only lay out the practices of dueling he witnessed in his career, but also to dig through history for every precedent and unusual incident he could find to offer an opinion on.

---

[38] Morgan 1ᵛ (pp 42-45).
[39] Latin: *Paridis* or *Paris de Puteo*.
[40] The edition I used is the first Italian edition: Paride del Pozzo. *De re militari*. Napoli: Sixtus Reissinger, 1476-7. There were various 16th century Italian reprints, generally titled some version of *Duello, libro de re*. An earlier Latin edition was published in the early 1470s, probably also by Reissinger, titled *De duello, vel De re militari in singulari certamine*. Because there are no page numbers in either 15th century edition, I will identify passages according to book and chapter numbers.

This treatise was groundbreaking and defined the conversation about dueling for the subsequent century—nearly every dueling commentary until the late 16th century frames its arguments to some extent as a response to his.

Pozzo's discussion of unarmored dueling is brief but unambiguous:[41]

> "It brings to mind two knights who, having wager of battle, obtained a field from a prince. They had decided an agreement to fight unarmed apart from swords and without any body armor, and with this, each one would show his mettle in defending their cause, and in order to defend their lives they would put themselves in a state such that each had the semblance of a raving dragon. Seeing this, on the appointed day [the prince] didn't want the battle to be made, seeing that it was more suitable for vile butchers than for valorous knights, and for this work and provision the prince was highly praised.
>
> "And in a similar case for the worthy prince, he wouldn't permit such a battle except with military weapons, fighting at least partially armored (and partially unarmored), since combat without all the armaments necessary for military exercise isn't pertinent to good knights.
>
> "Likewise, good knights in the field are wont on occasion, in similar endeavors, to exercise their valorous persons in order to show their strength and to defend their rights. And it's described in Lombard law that battle between knights shouldn't be made with sticks, nor with stones (unless there be contrary testimonies, then in that hour they should fight with sticks and shields in order to prove which of them had spoken the truth).
>
> "It happened that two German knights came into Italy to fight unarmed apart from swords and knives, and having obtained a free field, they gave notice to a judge, to whom many knights appealed that he not permit that one be made to cruelly kill. And the field was revoked by the prince, and the judge made harmony

Fig. 13: Sword against dagger
Ms. Ludwig XV 13, 19r (ca. 1400)
J. Paul Getty Museum

---

[41] Pozzo book 4, chapter 3. Translation is my own.

between them with excusatory words that should be said by the challenged party."

This establishes two facts: first, that unarmored dueling was not allowed by law or custom, which goes a long way toward explaining why there are so few accounts of such duels in historical records. Second, that some people *really* wanted to do it anyway.

But what are these knights to do if they want to pursue a dueling form that is opposed by the powers that be and contrary to the Law of War? There is a second practice that Pozzo condemns in very similar terms,[42] sometimes known as dueling *alla maccia* ("to the woods"):[43]

"One has been challenged and requested that he should look for a secure place and look for a judge in order to do battle to the doubtful extreme. When he can't find one, it's charged that he has to go and fight in a solitary place with his enemy (such as a field or wood) such that they wouldn't be separated or stopped since they're unseen.

"Some say that yes, they should go, because necessity makes many things lawful that were otherwise unlawful, and because the sword is judge and gives testimony that the one who returns from the battle without wounds demonstrates that he is victor, just as, on the other side, the one that were dead or gravely wounded would testify of being the loser. And in this way, the sentence of the battle can be reported without a judge because the wounds show themselves to be the judge.

"But on the contrary, it's responded that doing this for demonstration of truth would be against the order of knighthood, and against the entire constitution of arms, which wants battle to be celebrated in the presence of some prince and many knights, by whose determination justice would be remitted and not otherwise. And doing otherwise would be most indecent, outside of all military discipline, and would be a

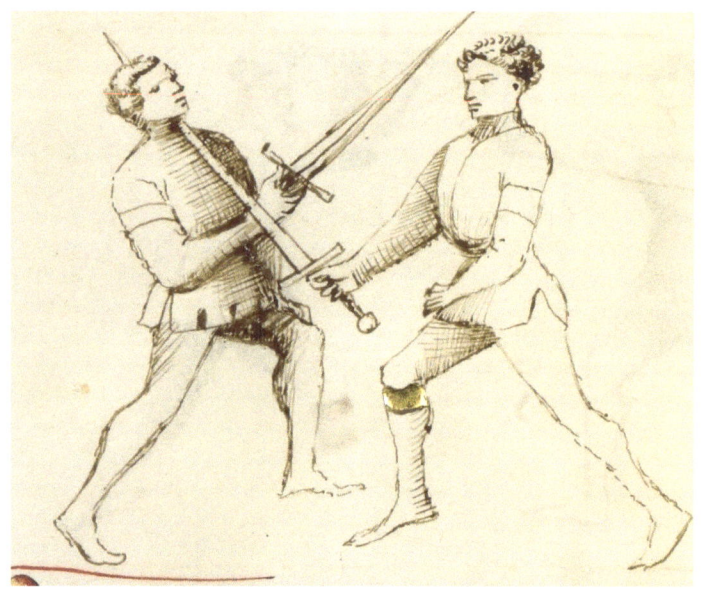

*Fig. 14: Unarmored fencing
Ms. Ludwig XV 13, 20ᵛ (ca. 1400)
J. Paul Getty Museum*

---

[42] Pozzo book 2, chapter 20. Translation is my own.
[43] HUGHES 105.

custom belonging to common butchers, ruffians, and hillbillies, which should be punished by judge and public justice. And things that aren't laudable shouldn't be used by knights, nor by other worthy men, for which it is said that the challenged knight doesn't have to go to the solitary place for the reason written above."

Fig. 15: A solitary place
Ms. Ludwig XV 13, 47ʳ (ca. 1400)
J. Paul Getty Museum

Another condemned practice, and Pozzo uses the same term of derision for both: suitable only for "butchers".[44] These are, in fact, the only two dueling practices that are described in these terms in the entire book, which seems to link them together.

Another thing that links them together is that both of these chapters have echoes of Fiore's own account: without armor, trusting only in his sword and his skills, he fought in strange places and the only outcome worth mentioning is that he escaped each duel without wounds. In the absence of new records about Fiore, I think this is the best explanation we're likely to get.

---

[44] *Beccharini.* This term is rendered "butcher" by 16th century translators, but also means things like "flesh-monger" and "grave-digger". The Latin edition instead says *lenonibus*, "pimps".

The Flower of Battle: MS M 383

Codicological Description

New York, The Morgan Library & Museum – s. xv^in – Venice.

Materials and physical description

Ms. M. 383 is a parchment manuscript with 20 leaves in folio format, measuring 277 × 195 mm. It was disbound in 2017 and currently has no cover or binding (fig. 1). The fore-edge shows evidence of gilding, but the top and bottom edges do not. The upper right corner of folio 19 was torn off and then patched (it's unclear when).

Fig. 1: The ms. neatly stacked Ms. M. 383, 1ʳ (ca. 1400) Morgan Library & Museum

Dating and origin

Estimates of the date of the manuscript typically start with the dedication of a different one: the *Flos Duellatorum* in the Museo archeologico Villa Pisani Dossi, also written by Fiore de'i Liberi. This manuscript contains the date 10 February 1409,[45] and that is used to estimate the date for all of Fiore's works. Thus, the Morgan Library offers a date of 1410-20 for their manuscript.[46]

There are various clues arguing for an earlier date, however. There's evidence in the composition of the manuscript itself that it predates the Pisani Dossi.[47] The Pisani Dossi also indicates that Fiore had been studying the art of fencing for 50 years,[48] whereas this manuscript says only that it had been more than 40 years.[49] A date between 1400 and 1410 seems most reasonable for the text, but if the illustrations were created first and text added later,[50] it could have been started decades earlier.

---

[45] NOVATI, carta 2^A.
[46] CORSAIR Holdings Information. Morgan Library & Museum. http://corsair.themorgan.org/vwebv/holdingsInfo?bibId=77302. Accessed 10 March 2021.
[47] See Appendix by LECCESE.
[48] NOVATI, carta 2^B.
[49] Morgan 2ʳ (pp 45-47).
[50] See Appendix by LECCESE.

The materials are hard to pin down to a specific region, but the language and artwork (see below) support an origin in Northern Italy, probably the Venetia region.

Provenance

The original owner of the manuscript is unclear; unlike the other two Italian manuscripts by Fiore de'i Liberi, it has no dedication to suggest a client. It may therefore have been his own personal copy.

The first known owner was Nicolò Marcello di Santa Marina († 1474) or heir, and his library seems to have remained intact in Venice for some time because it was from the Marcello library that Apostolo Zeno purchased the manuscript in 1699. Zeno produced a transcription of part of the preface to share with Giusto Fontanini (fig. 2), who was working on his seminal *Biblioteca dell'eloquenza italiana* ("Library of Italian Eloquence").[51]

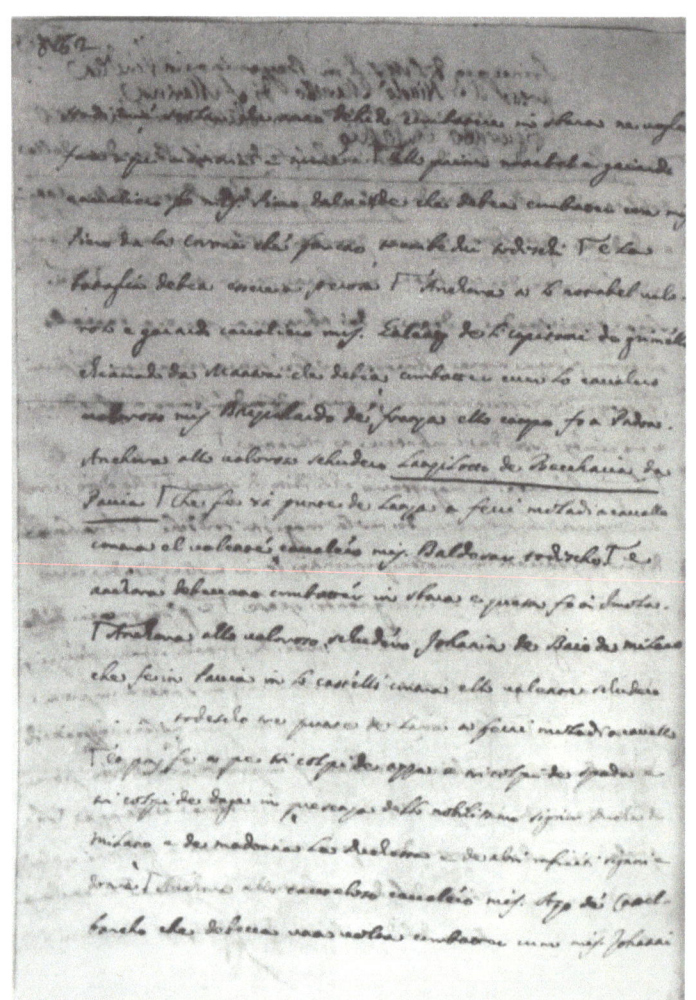

Fig. 2: Apostolo Zeno's copy Ms. XXIV, 783ʳ (1699) Biblioteca Guarneriana

From Zeno, it passed to the library of Venetian senator Jacopo Soranzo (1686–1761), where it eventually formed folia 241 to 260 of a larger compilation titled *Arte di armeggiare a piedi ed acavallo* ("Art of the Exercise Arms on Foot and on Horse", cod. MCCLXI).[52] (The other contents of this manuscript are unknown.) These folio numbers appear in the lower margin of the manuscript, and they show that the manuscript had already been cut into individual leaves by that time, and its original order had been disturbed. The manuscript had also lost some of its contents (see below) and been reduced to its current length of 20 folia.

The Soranzo Library was disbursed in 1780, and the manuscript was part of a large segment that entered the library of Matteo Luigi Canonici (1727–1806), another Venetian and a former Jesuit.[53]

---

[51] NOVATI 36-39.
[52] CORSAIR Holdings Information. Morgan Library & Museum. http://corsair.themorgan.org/vwebv/holdingsInfo?bibId=77302. Accessed 10 March 2021.
[53] Ibid.

In 1836, it was included in a group of manuscripts from the Canonici Library that was sold to Rev. Walter Sneyd (1809–1888) of Keele Hall, Staffordshire.[54] During his ownership, the manuscript was rebound in red morocco by J. Clarke and given the title *Trattato della scherma* ("Treatise on Fencing").[55] The inside cover bore his ex-libris and a sticker with the signature number 228.

Sneyd's collection returned to Sotheby's in a series of auctions in the early 20th century, and the manuscript was purchased by antiquarian and rare book dealer Tammaro de Marinis (1878–1969) in 1903.[56]

De Marinis sold the manuscript to John Pierpont Morgan (1837–1913) in 1909, and it has formed part of the Morgan Library collection ever since.[57] Morgan added his own bookplate to the inside of the front cover below Sneyd's, as well as the accession number "M. 383" in pencil. It was subsequently rebound by Marguerite Duprez Lahey in three-quarter tan Levant goatskin with block-printed paper, preserving the Clarke endpapers (fig. 3). The manuscript was most recently disbound in the Thaw Conservation Center in 2017 and remains unbound.[58] Its current call number is B1 361 B MS M.0383.

The inside cover of the Duprez Lahey binding bore an additional pencil note "V 4 B", but this can't currently be associated with a specific owner.

Fig. 3: The Duprez Lahey cover

Content

The manuscript's entire contents appear to be the work of Fiore de'i Liberi. It bears no title and is the only copy of his works that lacks

---

[54] According to a pencil note inside the front cover of the Duprez Lahey binding, it was auctioned by Sotheby's in London on 15 June 1836, no. 40.
[55] CORSAIR Holdings Information. Morgan Library & Museum. http://corsair.themorgan.org/vwebv/holdingsInfo?bibId=77302. Accessed 10 March 2021.
[56] According to a pencil note inside the front cover of the Duprez Lahey binding, it was auctioned by Sotheby's in London on 29 December 1903, no. 433.
[57] "De Marinis 1909" was written in pencil inside the back cover of the Duprez Lahey binding.
[58] CORSAIR Holdings Information. Morgan Library & Museum. http://corsair.themorgan.org/vwebv/holdingsInfo?bibId=77302. Accessed 10 March 2021.

The Flower of Battle: MS M 383

1^r – 2^r	Preface	
	Fiore furlan de Civida d'ostria che fo de mi<u>sser</u> Benedeto	
2^v – 7^r	Mounted fencing	
	Aquesto magistro ha ligada una forte corda alla sella dello suo cavallo	
7^v – 8^r	Mounted wrestling	
	Aquesto e zogo de abrazar e tanto, e a dire zogo de abrazar che zogo di braci	
8^{rv}	Javelin against cavalry	
	Aqui sono tri compagni che volemo ferire aquesto magistro	
9^{rv}	Lance	
	Aqui comenza larte de nobele arma chiamada lanza	
10^r – 11^v	Sword in armor	
	Aqui comenza la spada de armizare	
12^r – 16^v	Sword in two hands	
	*Aqui comenzano le guardie de spada a do mane e sono ·xii· **g**uardie*	
18^{rv}, 17^r	Dagger against sword	
	Aqui comenza spada e daga a zugare, lo avantazo si e gra<u>n</u>de a chi lo sa fare	
17^v, 19^r	Sword in one hand	
	Noii semo tri zugadori che volemo ferir questo magistro	
18^r	Sword against lance	
	Io aspetto aqui in dente de zenghiar, tuti tagli e punte io so riversar	
18^r	Spear and dagger against lance	
	Ad aquesto p<u>a</u>rtito si e de bastone, e daga <u>contra</u> lanza	
19^v – 20^v	Blank	

some version of the title "Flower of Battle"; by convention, the first line would be catalogued as the title, which is (perhaps intentionally) the author's name: *Fiore furlan de Civida d'ostria che fo de mi<u>sser</u> Benedeto*.

The table above lists the individual sections, including a transcription of the beginning of each section for reference. Apart from the preface, the format consists of 1-4 paragraphs of rhyming prose per page with an illustrative drawing below each one.

The preface mentions additional contents that are not present in the current manuscript, including teachings on the ax, wrestling, and the dagger.[59] Assuming the description of the manuscript contents are accurate, we can identify two lacunae: plays of the ax would fall between folia 16 and 17, and plays of wrestling and dagger would continue after folio 19. The ax plays could have been confined to a single missing leaf (for example, with the four guards and the four plays common across all extant copies occupying the recto and verso sides respectively) and the wrestling plays (which

---

[59] Morgan 2^r (pp 45-47).

are entirely consistent across the extant copies) would likely occupy two and a half folia, but there is too much variation in the dagger teachings to speculate about how long that section might have been.

Collation and foliation

As a collection of unbound single leaves, the manuscript has no quire structure and its collation formula would just be +20.[60] There's no way to determine whether it once consisted of bifolia or was a sheaf of individual leaves from the beginning, but its folia were in their current state no later than the 18th century when at least three leaves were bound out of order in Codex Soranzo.[61]

The fact that the manuscript was disordered in its 18th century binding lends credence to the idea that other leaves might also be out of their intended order. It's unclear which owner provided the current foliation in the upper right corners, but it was presumably after an attempt to order the leaves correctly when they were removed from Codex Soranzo and rebound. Based on analysis of their contents and comparison to other versions of the treatise, it seems likely folio 16 should come before 15 and folio 18 should come before 17. This consolidates the teachings on dagger against sword and sword in one hand into continuous sections and brings the manuscript in line with the description in the preface.[62]

Fig. 4: Damage to the inner edges of the three displaced folia Ms. M. 383, 14-16 (ca. 1400) Morgan Library & Museum

---

[60] The older library record gives a collation statement of $1^2 + 2^6 + 3^4 + 4^6 + 5^2$, but this seems to be erroneous; the record was written prior to the Duprez Lahey rebinding, so the collation must have been guessed at based on examining the spine of the Clarke binding. See CORSAIR Holdings Information. Morgan Library & Museum. http://corsair.themorgan.org/vwebv/holdingsInfo?bibId=77302. Accessed 10 March 2021.
[61] The numbering in the lower margin shows that folia 14-16 were bound in between folia 1 and 2.
[62] Morgan 2^r (pp 45-47).

In addition to the missing sections suggested above, comparison to other versions of the text suggest that two further leaves were lost: compared to other versions, exactly eight plays of the sword in two hands are omitted, which would fit sequentially between the final play on folio 14 and the first on 16, and eight plays of the sword in one hand are also omitted, which would likewise fit sequentially between the final play on folio 17 and the only play on 19.

The manuscript has three sets of folio numbers. In the lower margin are numbers from 241 to 260 in an 18th century hand, added when it was part of the larger Codex Soranzo. In the upper right corners are a set of small 18th century folio numbers from 1 to 20 with boxes drawn around them; these tend to disappear into the edges and were replaced by pencil numbers in the 19th century.

Writing and decoration

The manuscript is written in pale brown ink by a single hand in Italian Semigothic, a transitional hand between Gothic and Humanist bookhand which arose in Northern Italy in the mid to late 14th century. It combines characteristics of both scripts, and some of the distinctive letter forms are highlighted in figure 5, including the "biting" of round letters (*de, do, oc, oe*), the indiscriminate mixing of the standard *r* and the ⁊ or "round *r*" (which would only follow round letters in Gothic), and the distinctive *g* glyph and *st* ligature.[63] Another glyph distinctive to Semigothic that is used frequently in the manuscript (but not in figure 5) is the terminal 3 or "turned *m*", which is again mixed indiscriminately with the standard *m*.[64]

Fig. 5: Distinctive characters Ms. M. 383, 1r (ca. 1400) Morgan Library & Museum

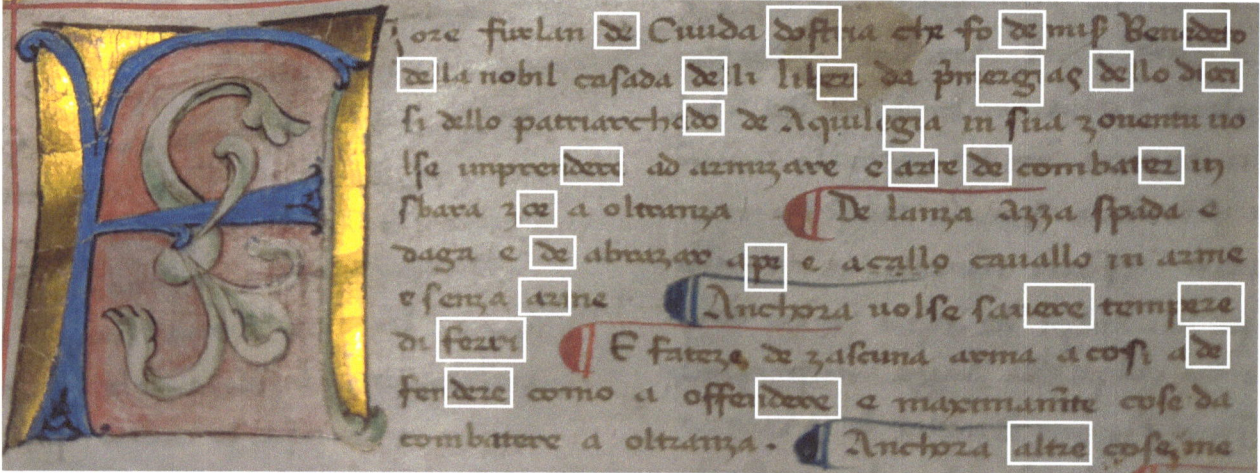

[63] CLEMENS AND GRAHAM 171-173.
[64] For example, in "*cu3*" on fol. 2r, col. 2, line 8. See pp 46-47.

The text is written in rhyming prose; for analysis of the language, I turn here to specialists. LEONI's description of the language in the Getty ms. is also apt for our manuscript:

"Spelling is non-standardized, as is to be expected in pre-1500 Italian, and it presents a fair amount of Lombardisms in the almost constant assibilation of consonants like *c* and *g* (example: *cioe'* ==> *zoe* or *çoe*) as well as the sonorization of the letter *t* (example: *dato* ==> *dado*) and the soft mutation of hard *c* into *g* (example: *gioco* ==> *zogho*). This places the manuscript and the scribe squarely in northeastern Italy, in an area that I would identify as being roughly between Friuli and Northern Emilia-Romagna."[65]

MALIPIERO expands upon this with an explanation of the Venetian dialectical markers specific to the Morgan:

"The text of the Morgan with respect to that of the Getty presents a very strong Venetian coloring: the prevalence of degeminate consonants (*tute, delo, fero, falita, butare, falimento* as opposed to the Getty's *tutte, dello, ferro, fallita, buttare, fallimento*), the frequent apocope of the infinitive (*pasar, falir, ferir* as opposed to the Getty's *pasare, fallire, ferire*), abbreviations like *savè* in place of the Getty's *savete*, the dropping of the intervocalic *g* and *l* (*maistro, meio, voio, taio,* where in the Getty we find *magistro, meglio, voglo, taglo*), and scattered other things like *tri* (always written *tre* in the Getty). The demonstratives and the adverbs of place in the Morgan are almost always characterized by prosthesis (*aquesto, aquello, acolui, acosì, aqui*) which is quite rare in the Getty (which almost always writes *questo, quello, choluy, così, qui*)."[66]

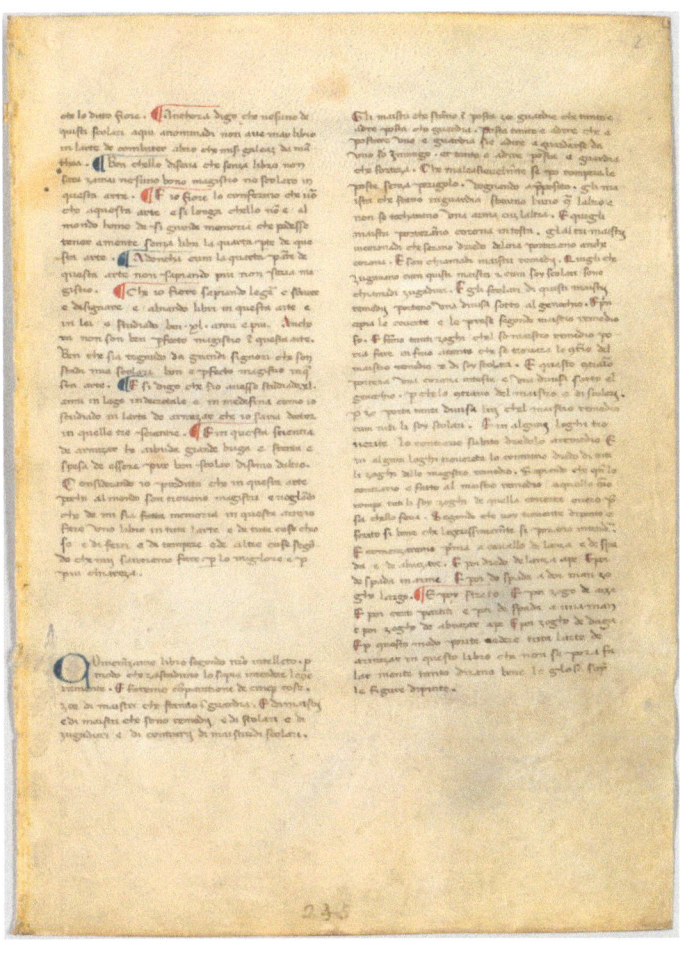

Fig. 6: Ms. M. 383, 2r (ca. 1400) Morgan Library & Museum

---

[65] LEONI AND MELE 173.
[66] MALIPIERO 130. "Il testo del Mrg rispetto a quello del Getty presenta una più forte coloritura veneziana: la prevalenza delle consonanti degeminate (*tute, delo, fero, falita, butare, falimento* di fronte a *tutte, dello, ferro, fallita, buttare, fallimento* del Getty), la frequentissima apocope dell'infinito (*pasar, falir, ferir* di fronte a *pasare, fallire, ferire* del Getty), abbreviazioni come *savè* in luogo di *savete* del Getty o la caduta della *g* e della *l* intervocalica *maistro, meio,*

Fig. 7: Spraywork border
Ms. Ludwig XV 13, 1r (ca. 1400)
J. Paul Getty Museum

Fig. 8 (op.): "De viris illustribus"
Ms. Latin 6069f, 1r (1379)
Bibliothèque nationale de France

The writing frame measures approximately 200 × 170 mm, but the ruling is mostly effaced and there are no examples where all four sides are visible. Folio 1r has a single column of text with 33 lines, whereas 1v-2r have two columns with 39-40 lines. The remainder of the manuscript is divided into four blocks per page, with one,[67] two,[68] or four[69] paragraphs of 3-13 lines and occasionally a 2-3 line paragraph above the frame occupying the full width.[70]

There is a single illuminated ten-line Lombard on 1r (fig. 1 & 5), and a two-line Lombard with faint pen flourishes on 2r (fig. 6). 1r has a full border outlined in red and blue ink, decorated with illuminated squares in the corners and four illuminated weapons: a dagger (top), a sword (left), an ax (right), and a lance (bottom). This has the appearance of an unfinished version of the spraywork border seen in Getty (fig. 7).

Illustrations

There are 108 delicate pen drawings in the manuscript by three artists spread across 2v-19r. This includes 92 confined to a single block and 16 that span two blocks. The figures within each block range in size from 130 to 380 mm wide and 215 to 200 mm tall.

The pen drawings are rather faint and some pages show evidence of redrawing.[71] The initials use blue, red, and yellow tempera colors in addition to gold leaf.

---

voio, taio, dove nel Getty troviamo *magistro, meglio, voglo, taglo*, o altri sparsi come *tri* sempre scritto *tre* nel Getty. Nel Mrg i demostrativi a gli avverbi di luogo sono quasi tutti carattarizzati da prostesi (*aquesto, aquello, acolui, acosì, aqui*) che è piuttosto rara nel Getty (che scrive quasi sempre *questo, quello, choluy, chosì, qui*)."

[67] 17r, 19r.
[68] 2v, 7r, 8v.
[69] 3, 4, 5, 6, 7v, 8r, 9, 10, 11, 12, 13, 14, 15, 16, 17v, 18.
[70] 9r, 10r, 13v, 16r.
[71] See Appendix by LECCESE.

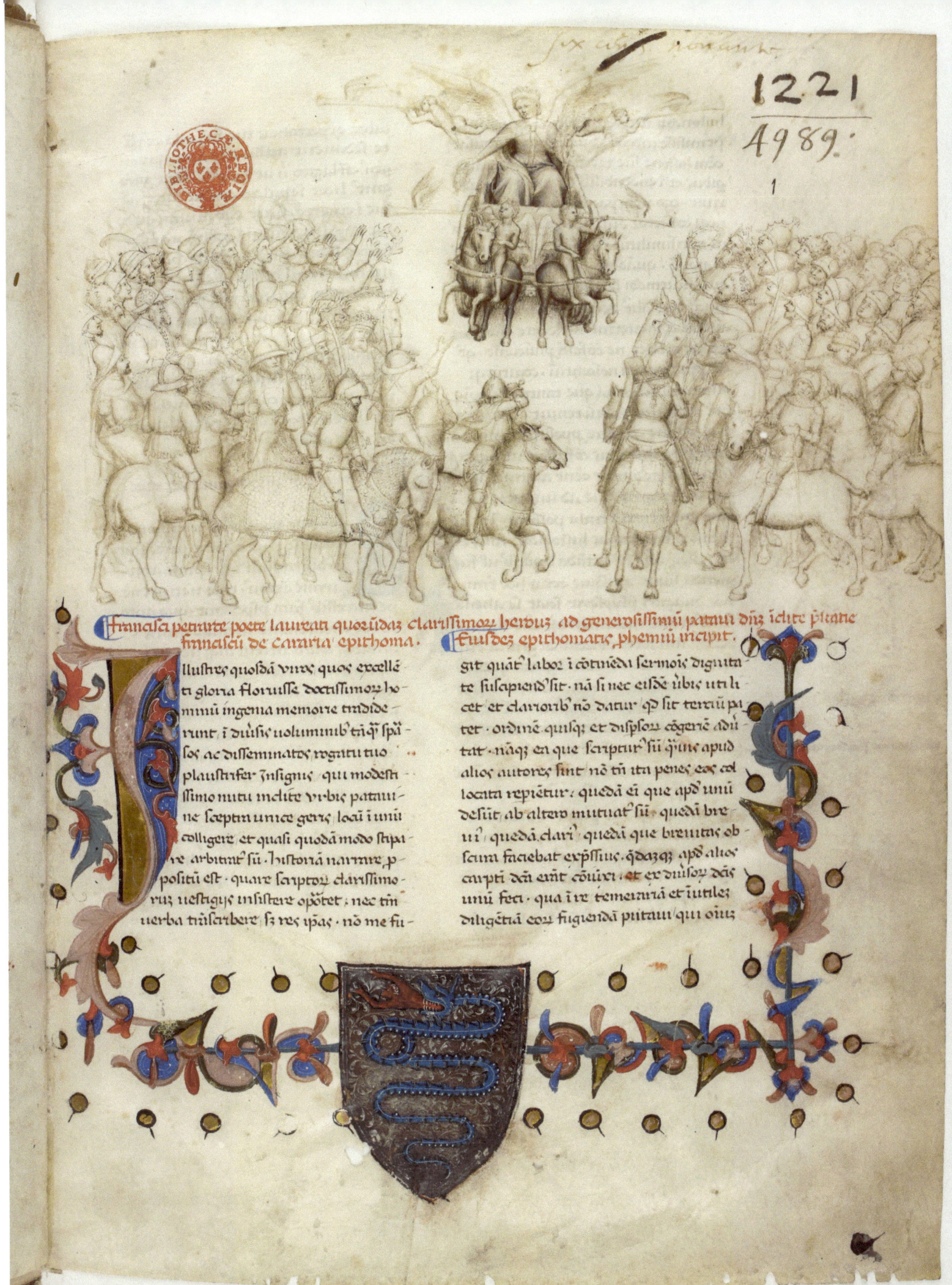

Francisci petrarce poete laureati quorundam clarissimorum heroum ad generosissimum patruum dnm iccnte p̄ntiāne
franscisci de cararia epithoma. Eiusdem epithomatis p̄hemium incipit.

Illustres quosdam viros quos excellenti gloria floruisse doctissimorum hominum ingenia memorie tradiderunt. i diuisis uoluminibu trī qñ spaciosos ac disseminatos rogantitus plaustrifer insignis, qui modestissimo nutu inclite vrbis pataunie sceptru unice geris, loca i unū colligere et quasi quodā modo stipare arbitrat̄ sū. historiā narrare ppositū est. quare scriptor clarissimorum uestigiis insistere opotet. nec tn uerba trāscribere si res ipsas. nō me fu-

git quāt labor i cōtinēdi sermois dignitate suscipiendo sit. nā si nec eidē ubis uti licet et clariorib' nō datur q sit terciū patet. ordine quisq; et dispoʒ cōgerie aduitat. nāq; ea que scripturū sū ī inue apud alios autores sint. nō tn ita penes. eos collocata repiētur. quedā ei que apd uniū desūt. ab altero mutuat' sū. quedā breuiū. quedā clari' quedā que breuitas obscurū faciebat expssius. qdāq; apd alios carpti dā eunt cōuici. et ex diuisoʒ dās unū feci. quā i re temerariā et inutile; diligētiā eoʒ fugiendā putaui qui oiu

The artists were Venetian or Veronese and may have been followers of Altichiero da Zevio.[72] An art-historical analysis comparing Altichiero's works to Fiore's manuscripts is beyond the scope of this work, but two representative samples are included on the facing page (figs. 8 & 9). This merits future specific study.

Artist A contributed nearly all of the illustrations, whereas Artist B's work is only seen on 17ᵛ and 19ʳ; artist B likely also worked on the Getty or was a follower of the Getty artist. Artist C did the bold redrawing apparent especially on 10ʳ, as well as more minor redrawing on 4ʳ, 6ʳᵛ, 7ʳᵛ, 11ᵛ, 13ʳ, 14ᵛ, 15ʳᵛ, 16ᵛ, and 18ᵛ; artist C is identified by LECESSE with the Pisani Dossi artist.[73]

As Fiore explains in the preface, the masters and students bear tokens to differentiate them from each other and the "players" or "companions": the masters wear crowns and the students wear a garter (usually under one knee), while the counter masters wear both crown and garter. These tokens are almost always leafed with gold (fig. 10), though a few were missed.[74]

A unique aspect of this manuscript is that the fencer who has the advantage in each exchange is further indicated by silver leaf on the metallic portions of their weapon (in case of an equal engagement, both blades are silvered). The silver leaf has become heavily tarnished over the centuries and is now varying shades of dark grey; it appears black under direct light, but indirect light or an oblique viewing angle still produces a silver sheen (fig. 10). Unfortunately, due to the buckling of

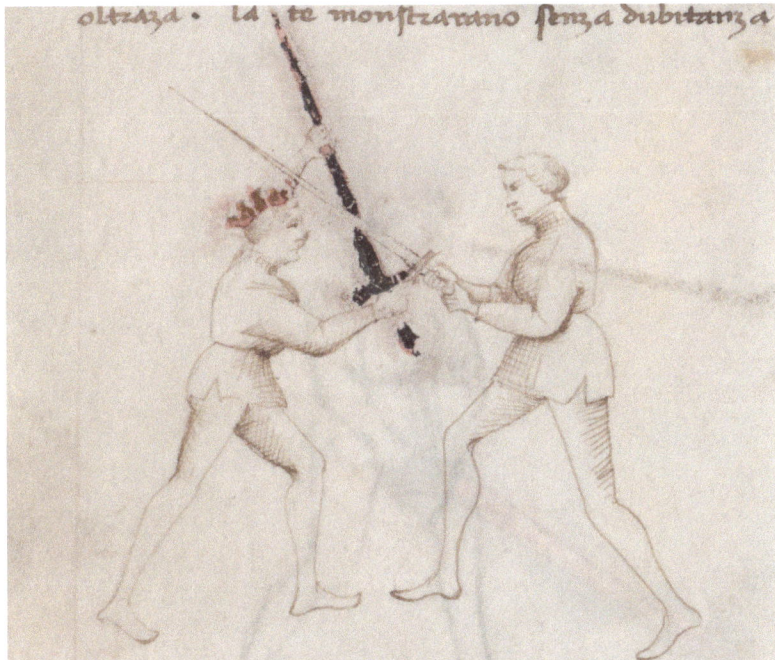

Fig. 10: Lighting the metallic leaf
Ms. M. 383, 10ᵛ (ca. 1400)
Morgan Library & Museum

Fig. 9 (op.): "Conquest of a Town"
National Gallery of Art

---

[72] ПАЛІЙ 88. MELLINI states that 2ᵛ-8ᵛ were drawn by Altichiero himself and ascribes 9ʳ-19ʳ to a follower. He offers no justification for this statement, and on the latter point might have been deceived by the same redrawing noted by LECESSE. See MELLINI 10-11 and Appendix by LECCESE.
[73] See Appendix by LECCESE.
[74] The counter master on 7ᵛB (pp 68-69), the student on 11ʳD (pp 82-83), both counter masters on 13ᵛ (pp 92-93).

the parchment it's nearly impossible to illuminate all the metallic elements on a leaf at once (fig. 11).

If LECESSE's hypothesis is correct that carta 13 from the NOVATI facsimile represents a leaf removed from this manuscript and later replaced on 17ᵛ and 19ʳ by figures based on the Getty, then the silver leaf was added to the pen drawings much later, after the Getty and Pisani Dossi manuscripts were completed.[75]

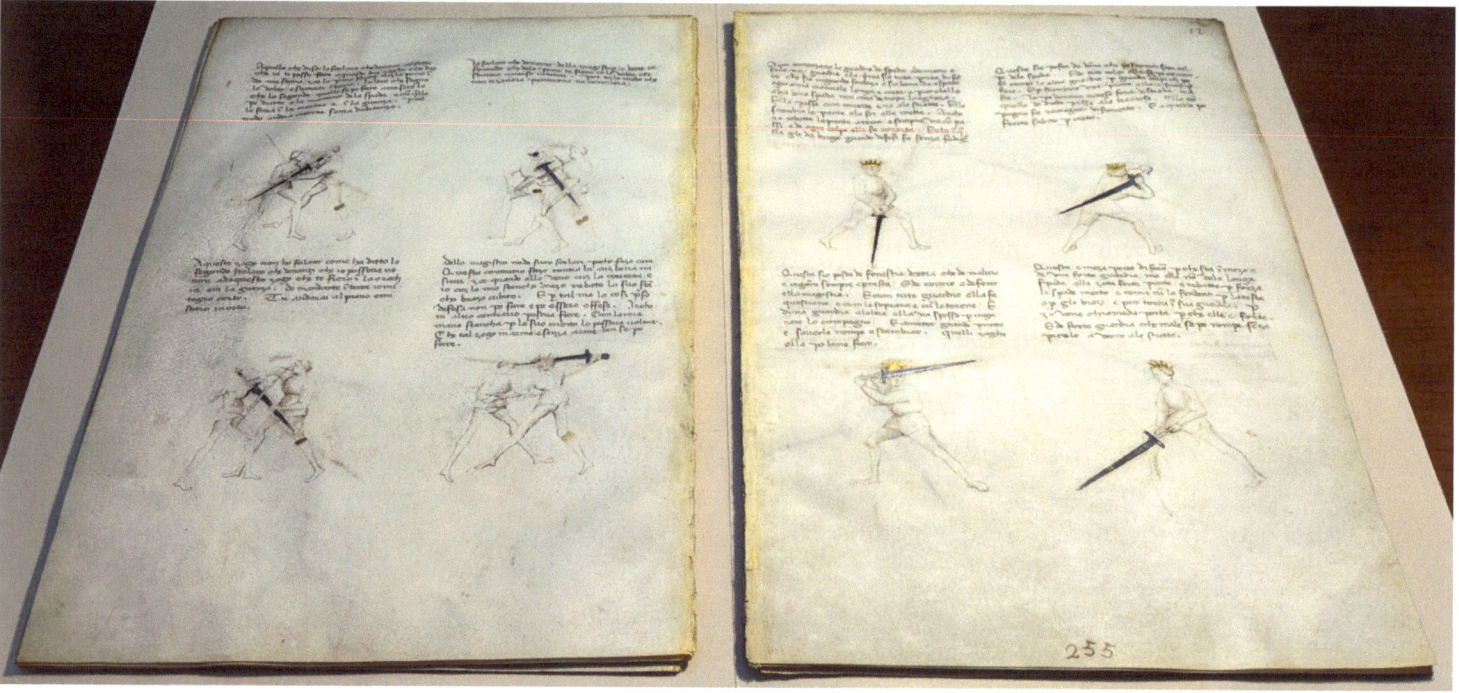

Fig. 11: A two-page spread
Ms. M. 383, 11ᵛ-12ʳ (ca. 1400)
Morgan Library & Museum

---

[75] See Appendix by LECCESE.

Transcription and Translation

The layout of this section is straightforward: a page scan reproduced at its real size on one side, and the transcription and translation on the other; raising a translation page vertical will allow you to view the two-page spread. The pages are arranged according to the speculative original order outlined earlier (rather than the current one).

It's impossible to transcribe a manuscript into print and perfectly represent the text on the page, and I haven't tried to do so. I have, however, tried to stay close to the text and avoid injecting 21st century ideas about spelling or punctuation into it.

I also haven't represented the line breaks in this transcription, for the simple reason that the script is too small and doing so in this format would require print that was too small for easy reading.

Where variant glyphs are used, I've transcribed them according to the modern form: ʒ becomes *m*, ꝛ becomes *r,* and ſ becomes *s*.

Abbreviations are common in this manuscript, and I've resolved them according to accepted paleographical standards with the omitted letters <u>underlined</u>. The more common abbreviations are given in the sidebar.

I've rendered virgules as commas and terminal marks as periods; there are potential diacritics in the manuscript, but they are too inconsistent and too similar to stray marks, so I haven't tried to include them. *Capitulum* marks are represented by pilcrows in the appropriate color (¶), and letters that have been struck with red ink are represented with leading red pipes (|). Deletions are indicated with ~~strikethrough~~, and cases of obvious scribal error are indicated with [!].

A translator must always find a balance between trying to capture the voice of the author and trying to capture his language. Fiore's authorial voice is informal and conversational. Indeed, it almost has the character of a dictation, full of rambling sentences, verbal stutters, and self-corrections. His language, on the other hand, is a Northern Italian dialect that is rather far removed from English, written in broken rhyming prose.

In preparing this translation, I've leaned more in the direction of capturing Fiore's words than his tone, though I believe I've done a fair job of capturing both. I want the translation to be easily compared with the transcription, so while my translation is hardly word-for-word, I've tried to keep his ideas and clauses in roughly the same order and not paraphrase too heavily. I have not, however, attempted to preserve the rhyme scheme.

In places where my translation deviates too far from what a fencer might expect, I've included footnotes explaining my choices.

Abbreviations

 *de*

 *par*
per
por

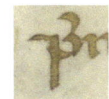 *pra*
pre
pro

 *sser*

 *con*

 *& (et)*

 (vowel)
(e)*m*
(e)*n*

 *r*
r with vowel

1ʳ *FIore furlan de Civida dostria che fo de misser Benedeto della nobil casada delli liberi da premergiaco dello diocisi dello patriarchado de Aquilegia in sua zoventu volse imprendere ad armizare, e arte de combater in sbara zoe a oltranza, ¶ De lanza, azza spada, e daga, e de abrazar a pe, e a ~~callo~~ cavallo in arme e senza arme. ¶ Anchora volse savere tempere di ferri. ¶ E fateza de zascuna arma, a cosi a defendere como, a offendere, e maximamente cose da combatere, a oltranza. ¶ Anchora altre cose mereveglose, e occulte che a pochi homeni del mondo sono palese. ¶ E sono cose verissime, e de grandissima offesa, e deffesa, e cose che non se pon falare tanto sono lizere a fare, ¶ la quale arte, e magisterio che ditto di sopra. ¶ E llo ditto fiore si a imprese le ditte cose de molti magistri todeschi, ¶ Anchora de molti ijtaliani in molti provintie, e in molte zitade, cum grandissima fadiga, e cum grande spese, ¶ E per la gracia de dio de tanti magistri e scolari. ¶ E in corte di grandi signori, principi duchi Marchesi, e Conti, cavalieri, e scuderi in tanto, a impresa aquesta arte, ¶ Che llo ditto fiore i estado piu, e piu volte requirido da molti signori e cavaleri, e scuderi per imprendere dal ditto fiore si fatta arte, de armizar e de combatere in sbara ¶ a oltranza, la quale arte ello a mostrada, a piu sori ijtaliani e todeschi e altri grandi signori che anno debudo combatere in sbara. ¶ E anchora, a Infiniti che non anno debudo combater. ¶ E de alguni che sono stadi mie scolari che anno debudo combatere in sbara, ¶ Di alguni ne voglo fare aqui memoria, e nome ¶ E llo primo notabel, e gaiardo cavaliero fo misser Piero dal verde che debea combater cum miser Piero dala corona che foreno trambedui todeschi. ¶ E la batagla debea essere a perosa. ¶ Anchora allo valoroso cavaliero miser Nicholo unriçilino todescho che debea combater cum nicholo Inghileso e llo campo fo dado a Imola. ¶ Anchora alo notabel ~~va~~ valoroso, e gaiardo cavalero misser Galeaz delli captani de grimello chiamado da Mantoa che debea combater cum lo cavalero valoroso miser Buçichardo de franza, e llo campo fo a padoa.*

In his youth, Fiore Friulano de Cividale del Friuli, son of Sir Benedetto of the noble house of the Liberi of Premariacco in the diocese of the Patriarchate of Aquileia, wanted to learn fencing and the art of combat in the list (that is, in the extreme) with lance, ax, sword, and dagger, and of wrestling, on foot and on horse, in armor and without armor.

He also wanted to know about the temper of iron, and the qualities of each weapon, as much for defense as for offense, and most of all matters of combat in the extreme.

Also other marvelous and occult things that are apparent to few men in the world, and are very true things and very great for offense and defense, and things that cannot fail you since they are so easy to do, which art and mystery is mentioned above.

And said Fiore did learn said things from many German masters and from many Italians too, in many provinces and in many cities, with the greatest fatigue and with great expense, and (by the grace of God) from so many masters and students.

And he undertook this art in the courts of great lords, princes, dukes, marquises, counts, knights, and squires, and so said Fiore was more and more often retained by many lords and knights and squires in order to learn from said Fiore how to do the art of fencing and of combat in the list in the extreme. He demonstrated this art to many Italians and Germans and other great lords that were obliged to combat in the list (and also to countless that were not obliged to combat).

And for some who've been my students and have been obliged to combat in the list, I want to name them and make a remembrance here.

And the first notable and gallant knight was Sir Peter von Grünen, who was obliged to combat with Sir Peter Kornwald (who were both Germans). And the duel was required to be at Perugia.

Also the valiant knight Sir Nikolo Unriçilino (a German), who was obliged to combat with Nicholas (an Englishman), and the field was given at Imola.

Also the notable, valiant, and gallant knight Sir Galeazzo Cattaneo dei Grumelli, called da Mantua, who was obliged to combat with the valiant knight Sir Boucicaut, [Marshal] of France[76], and the field was at Padua.

---

[76] Jean II le Maingre.

Anchora allo valoroso schudero lanzilotto de bocharia da pavia. Che fe vi. punte de lanza a ferri moladi a cavallo, contra el valente cavalero miss. Baldesar todescho. E anchora debevano combater in sbaria, e questo fo a Jmola. Anchora allo valeroso schudero Zohanni de Baio da Milano che fe in pavia in lo castello, contra ello valente schudero Sram todescho tre punte de lanza a ferri moladi a cavallo. E poy se a pe tri colpi de azza, e tri colpi de spada e tri colpi de daga, in presenza dello nobilissimo signor ducha de Milano e de madona la duchessa e de altri infiniti signori e done. Anchora allo cauteloso cavalero miser Azo da Castelbarcho che debeva una volta combater cum miss. Zohanni di li ordelaffig. E una altra volta cum ello valente e bono cavalero miss. Jacomo da Boson el campo debeva essere allo piaso del signor ducha de milano de questi e de altri che io fiore ho amagistradi e sono molto contento per che son stado bene remunerado, e sio aibudo lo honore ello amore di mie scolari e de loro parenti. Anchora digo che achi io insignada questa arte. Jo lo insignada occultamente che llo non gle stado persona altra che lo scolare e algino districto suo o parente. Anchora che aquilli che gli sono stadi aibudo sacramento de non apalentar nesuno zogho che loro abiano veçudo da mi fiore. E maximamente me lo guardado da magistri scurmiduri e de soi scolari. E loro per invidia zoe li magistri mano convidado a zugar a spade di taglio e de punti in zuparello da armare senza altra arma salvo che un pazo de guanti de camoza, e tuto questo e stado per che io non ho vogludo praticar

cum loro, ne ho voglud insignare niente de mia arte. E questo accident e stado v. volte che sono stado requirido. E v. volte per mio honore ma convegnudo zugar in loghi strani senza parenti e senza amisi, non abiando speranza in altro. Che in dio, in larte e in mi fiore, et la mia spada. E per la gratia de dio io fiore sono romaso cu honore, e senza lesione de mia persona. Anchora io fiore disevo a mie scolari che debevan combatere in sbara, che lo combater in sbara e asa asa de meno periculo, che a combater cum spade de taglio, e de punta in zuparelo da Armar. Per che aquello che zoga a spade taiente, una sola coverta che falla a quello colpo gli da la morte. E uno che combate in sbara, e bene armato, e po recevere ferde asai anchora po vincer la batagla. Anchora sie una altra cosa che rare volte ne perisse nesuno per che se piglano a presone. Si che io digo voria umanca combater tre volte in sbara, che una sola volta a spade taiente, come dito de sopra. E si digo che lo homo che de combater in sbara siando bene azmato e servando larte de lo combater, e abiando li avantazi che se pon piglare se ello non e valente ello se voria ben impichare ben che posso dire per la gratia de dio che zamay nesuno mio scholaro in questa arte non fo perdidore. Che sempre sono romasi cum honore in questa arte. Anchora digo io predicto che aquisti signori cavaleri e schuderi che io monstrada aquesta arte da combatere sono stadi contenti delo mio insognare non voglando altro magio

1ᵛA /Anchora allo valoroso schudero lanzilotto de Beccharia da pavia, ¶ Che fe ·vi· punte de lanza a ferri moladi a cavallo, contra el valente cavalero misser Baldesar todescho, ¶ /E anchora debevano combater in sbarra, e questo fo a Imola. ¶ Anchora allo valoroso schudero Zohannj de Baio da Milano che fe in pavia in lo castello, contra ello valente schudero Sram todescho tre punte de lanza a ferri moladi, a cavallo ¶ /E poij fe a pe tri colpi de azza, e tri colpi de spada e tri colpi de daga, im[!] presenza dello nobilissimo signor ducha de Milano, e de madona la duchessa e de altri infiniti signori e donne. ¶ Anchora allo cauteloso cavalero misser Azo di Castelbarcho che debeva una volta combater cum misser Zohanni di li ordelaffig. ¶ /E una altra volta cum ello valente e bono cavalero misser Jacomo da Besen, el campo debeva essere allo piaser del signor ducha de milano, de questi e de altri che io fiore hoe amagistradi, e sono molto contento perche son stado bene remunerado, e si o aibudo lo honore, e llo amore di mie scolari e de lor parenti. /Anchora digo che a chi io insignada aquesta arte, Io lo insignada occultamente chello non gle stado persona altra che lo scolare, e alguno di stretto suo parente. Anchora che aquilli che gli sono stadi anno aibudo sacramento de non apalentar nesuno zogho che loro abiano vezudo da mi fiore. ¶ E maximamente, me ho guardado da magistri scrimiduri e de soi scolari. E loro per invidia zoe li magistri mano convidado a zugar a spade de taglio e de punta in zuparello da armare senza altra arma salvo che un paro de guanti de camoza, e tuto questo e stado perche io non ho vogludo praticar

Also the valiant squire Lancillotto Beccaria of Pavia, who made 6 thrusts of sharpened iron lance on horseback against the valiant knight Sir Balthasar von Braunschweig-Grubenhagen (a German). They were obliged to combat in the list, too, and this was at Imola.

Also the valiant squire Giovannino da Baggio of Milan, who made three thrusts of sharpened iron lance on horseback against the valiant squire Sirano (a German) in the castle in Pavia. And then, dismounted, he made three blows of ax, and three blows of sword, and three blows of dagger, in the presence of the most noble lord Duke of Mi-lan, and of the lady Duchess, and of countless other lords and ladies.

Also the cautious knight Sir Azzone di Castelbarco, who once was obliged to combat with Sir Giovanni di Ordelaffi, and another time with the valiant and virtuous knight Sir Giacomo da Boson (the field to be set at the pleasure of the lord Duke of Milan).

I, Fiore, have taught these and others, and I'm very content because I've been compensated well and I've had the honor and love of my students and their relatives.

Also I say that to whomever I taught this art, I've taught it secretly so there were no person other than the student and some close relative of his. Also that those who were present had sworn on the sacrament that they wouldn't reveal any play that they'd seen from me, Fiore.

And most of all, I've been wary of fencing masters and their students. And out of envy, they (that is, the masters) challenged me to play at swords of sharpened edge and point, in arming jacket but without any other armor save for a pair of chamois gloves. And all this just because I didn't want to practice

1ᵛB *cum loro, ne ho vogludo insignare niente de mia arte. ¶ E questo accident e stado ·v· volte che sono stado requirido. ¶ E ·v· volte per mio honore ma convegnudo zugar in loghi stranij senza parenti, e senza amisi, non abiando speranza in altro, /Che in dio, in larte e in mi fiore, e in la mia spada. ¶ E per la gratia de dio io fiore sono romaso cum honore, e senza lesione de mia persona. ¶ Anchora io fiore diseva a mie scolari che debevan combatere in sbara, che lo combater in sbara, a asa asa de meno periculo, che, a combater cum spade de taglio, e de punta in zuparelo da Armar. ¶ Perche aquello che zoga a spade taienti, una sola coverta che falla aquello colpo gli da la morte. ¶ /E uno che combate in sbara, e bene armato, e po recevere feride asai anchora po vincer la batagla. ¶ Anchora si e una altra cosa che rare volte ne perisse nesuno perche se piglano a presone. ¶ Si che io digo voria innanci combater tre volte in ssbara[!], che una sola volta a spade taiente come dito de sopra. ¶ E si digo che lo homo che de combater in sbara siando bene armato e sapiando larte delo combater, e abiando li avantazi che se pon piglare se ello non e valente, ello se voria ben impichare ben che posso dire per la gratia de dio che zamaij nesuno mio scholaro in questa arte non fo perdidore /Che sempre sono romasi cum honore in questa arte. ¶ Anchora digo io predeto che aquisti signori cavaleri e schuderi che io mon-strada aquesta arte da combatere sono stadi contenti delo mio insegnare non vogl-ando altro magistro*

with them, nor did I want to teach them anything of my art.

And this incident, that I was so challenged, occurred 5 times. And 5 times I had to play for my honor in strange places, without relatives and without friends, having no hope in anything other than in God, in the art, and in myself, Fiore, and in my sword. And by the grace of God, I, Fiore, remained with honor and without wounds on my body.

Also I, Fiore, told my students who were obliged to combat in the list that combat in the list is a far lesser peril than combat with swords of sharp edge and point in arming jackets.

Because for him that plays at sharp swords, if he fails even a single cover, that blow gives him death. But one that fights in the list and is well-armored can receive many strikes and can still win the duel. Here's another thing, too: that only on rare occasions does someone perish because of grips and holds.

Thus I say that I would sooner fight in the list three times than just once with sharp swords, as I said above.

And I say that a man, being well-armored for combat in the list, and knowing the art of combat, and having all the advantages that he can take, even so, if he isn't valiant then he'll wish to hang himself. Well can I say that, for the grace of God, none of my students in this art have been losers—that always they remained with honor in this art.

I also say that I predict that these lords, knights, and squires to whom I've demonstrated this art of combat are content with my teachings, not wanting any other master

2ʳA *che lo ditto fiore. ¶ Anchora digo che nesuno de quisti scolari aqui anominadi non ave maij libro in larte de combater altro che misser galeaz di manthoa. ¶ Ben chello diseva che senza libro non sara zamai nesuno bono magistro ne scolaro in questa arte. ¶ /E io /Fiore lo confermo che vero che aquesta arte e si longa chello non e al mondo homo de si grande memoria che podesse tener a mente senza libri la quarta parte de questa arte. ¶ /Adoncha cum la quarta parte de questa arte non sapiando piu non seria magistro. ¶ Che io fiore sapiando legere, e scriver e disignare, e abiando libri in questa arte, e in lei, o studiado ben ·xl· anni e piu. /Anchova[!] non son ben perfecto magistro in questa arte. Ben che sia tegnudo da grandi signori che son stadi mie scolari, bon e perfecto magistro in questa arte. ¶ /E si digo che sio avesse studiado ·xl· in lege in decretale, e in medesina como io studiado in larte de armizar che io saria doctor in quelle tre scientie. ¶ /E in questa scientia de armizar ho aibuda grande briga, e stenta, e spesa de essere pur bon scolare, disemo daltro. /Considerando io preditto che in questa arte pochi al mondo sen trovano magistri, e voglando che de mi sia fatta memoria in questa arte, io faro uno libro in tuta larte, e de tute cose chio so, e di ferri e de tempere e de altre cose segondo che nui saveremo fare per lo migliore, e per piu chiareza.*

COmennzamo libro segondo vero intelleto, per modo che zascaduno lo sapia intendere lezieramente. /E faremo comparatione de cinque cose, zoe di maistri che stanno in guardia. /E di maistri[77] che sono remedij, e di scolari, e di zugaduri, e di contrarij di maistri, & di scolari.

than the aforesaid Fiore.

I say as well that none of these students named here had any book about the art of combat other than Sir Galeazzo di Mantua. Well did he say that no one will ever be a good master (nor even a good student) in this art without books.

And I, Fiore, confirm that it's true that this art is so long that there's no man in the world with such a great memory that he can hold in his mind even a fourth part of this art without books, and I grant that if I didn't know more than a fourth part of this art, I wouldn't be a master.

Thus I, Fiore, knowing how to read and to write and to draw, and having books on this art, and having studied it for 40 years and more, yet I'm not a very perfect master in this art. (Though I'm held, by the great lords that have been my students, to be a good and perfect master in this art.) And I say that if I'd studied civil law, canon law, and medicine for 40 years, as I've studied in the art of fencing, then I'd be a doctor in those three sciences. But in this science of fencing, I've had great contentions and strain and expense just to be a good student (as we said of others).

Considering that I could find few masters of this art in the world, as I said earlier, and wishing that I might be remembered in this art, I'll put all the art into a book (and all things that I know of iron and of temper), and of other things following that which we know how to do for the best, and for the most clarity.

We begin this book following our intellect, in such a way that anyone would know how to understand easily.

And we'll make comparison of five things, that is, of masters who stand in guard, and of masters who are remedy, and of students, and of players, and of counters to masters and students.

---

[77] "e di maistri" appears twice in a row in the text, but isn't struck out like other duplications.

2ʳB /Gli maistri che sta*nn*o *in* posta zoe guardie che tanto, e a dre[!] posta che guardia. /Posta tanto e a dire che a postare uno, e guardia si e a dire a guardarse da uno so Inimigo, et tanto e a dire posta e guardia che forteza. /Che malea-sievelme*nt*e se po rompere le poste senza perigolo, Vegnando a p*ro*posito, gli magistri che stano in guardia, stanno luno *contra* laltro, e non se tocharano una arma cum laltra. /E quigli maistri portera*nn*o corona in testa, gli altri maistri incoronadi che serano driedo de lore porterano anche corona. /E son chiamadi maistri remedij. /Quigli che Zugarano cum quisti maistri & cum soij scolari, sono chiamadi zugaduri. /E gli scolari di quisti maistri remedij, portano una divisa sotto al genochio. /E p*ri*ncipia le coverte, e le prese segondo maistro remedio so. /E fa*ra*no tanti zoghi chel so maistro remedio poria fare in fino a tanta che se trovara lo *contra*rio del maistro remedio & di soij scolari. /E questo *contra*rio portera una corona in testa, e una divisa sotto el genochio, *per*che lo *con*trario del maistro e di scolari, *per*zo porta tanta divisa luij che lmaistro remedio cum tuti li soij scolari. /E in alguni⁷⁸ loghi troverite, lo contrario subito dredo lo arem-edio. /E in alguni loghi troverete lo contrario dredo di tuti li zoghi dello magistro remedio. Sapiando che qu*el*o contrario e fatto al maistro remedio, aquello *contra*rio rompe tuti li soij zoghi de quella coverta overo p*re*sa chello feva. /Segondo che voij troverite dipento e scrito si bene che leçerissimame*nt*e si porano intende*re*. /E comenzaremo p*ri*ma a cavallo de lanza, e de spada, e de abrazare. /E poi dredo de lanza a pe, /E poi de spada in arme, /E poi de spada a doi man zogho largo. ¶ E poij streto E poi zogo de azza /E poi certi partiti, e poi de spada a una man a poi zogho de abrazar a pe /E poi zogho de daga. /E *per* questo modo porite vedere tuta larte de armizar in questo libro che non se pora falar niente tanto dirano bene le glose sop*ra* le figure dipinte.

The masters are in positions (that is, guards—that which are called both positions and guards). Positions are called that because they position one, and guards are called such because they guard one from an enemy, and as such, they are called strong positions and guards. Poorly will anyone be able to break the positions without danger.

Coming to the point of this, the masters that stand in guard stand one against another without touching one weapon against the other, and here the masters will bear crowns on their heads.

The other crowned masters that will be after them also bear crowns, and they are called Masters Remedy.

Those that play here with these masters and with their students are called Players.

And the students of these Masters Remedy bear a token under their knees, and initiate the covers and holds following what the Master Remedy does, and doing such plays that the Master Remedy knows how to do.

At the end will be found the counter to the Master Remedy and to his students. And this counter bears a crown on his head and a token under his knee because he's the counter of the master and of the students, and as such he bears the tokens of both the Master Remedy and all his students. In some plays the counter will be found immediately after the remedy, and in some plays the counter will be found after all the plays of the Master Remedy. Know here that the counter that's made to the Master Remedy, that counter breaks all the plays of the cover or hold that he makes.

In the following, you'll find them drawn and written well so that they can be easily understood.

And we'll begin first on horse with lance, and with sword, and with wrestling. Then, afterward, with lance on foot, and then with sword in armor. Then with sword in two hands in loose play, and then tight [play]. Then the play of the ax, and then certain special matches, and then sword in one hand. Finally, wrestling on foot and then the play of the dagger.

In this way, you can see all the art of fencing in this book, and the explanations about the drawn figures are worded so well that it can never fail you.

---

⁷⁸ Corrected from "a"(?) to "i".

che lo ditto fiore. ¶ Anchora digo che nesuno de quisti scolari aqui anominadi non aue may libro in l'arte de combater altro che miss. galeaz da mantua. ¶ Ben ch'ello disena che senza libro non sara zamai nesuno bono magistro ne scolaro in questa arte. ¶ E io fiore lo confermo che no, che aquesta arte e sì longa ch'ello no e al mondo homo de sì grande memoria che podesse tenere amente senza libri la quarta p[ar]te de questa arte. ¶ Adoncha cum la quarta parte de questa arte non sapiando piu non seria magistro. ¶ Che io fiore sapiando lege[re] e scriuere e designare, e abiando libri in questa arte e in lei o studiado ben .xl. anni e piu. Anchora non son ben perfecto magistro in questa arte. Ben che sia tegnudo da grandi signori che son stadi mie scolari bon e perfecto magistro in sta arte. ¶ E sì digo che s'io auesse studiado .xl. anni in lege in decetale e in medesina como io studiado in l'arte de armizar che io saria doctor in quelle tre scientie. ¶ E in questa scientia de armizar ho abuda grande briga e stenta e spesa de essere pur bon scolar. Non stimo d'altro.

Considerando io p[re]ditto che in questa arte pochi al mondo son trouati magistri. E uoglando che de mi sia fatta memoria in questa arte o fato uno libro in tuta l'arte e de tuta cosa ch'io so e di ferri e de tempere e de altre cose segondo che mi sauerem[o] fare per lo migliore e per piu chiareza.

¶ Començamo libro segondo n[ost]ro intelecto. p[er] modo che zaschaduno lo sapia intendere e lege[re] uera mente. ¶ E faremo comparatione de cinque cose. ço è di maistri che stanno in guardia. E di maistri e di maistri che sono remedy. E di scolari e di zugaduri. E di contrary di maistri e di scolari.

Gli maistri che stanno in poste zoe guardie che tanto e adire posta che guardia. Posta tanto e adire che a postare uno e guardia sic adire a guardarse da uno so inimigo. Et tanto e adire posta e guardia che fortezza. Che malesseuelmente se p[ò] rompere le poste senza perigolo, vegnando apposito. Gli maistri che stanno in guardia stanno l'uno e l'altro e non so tochan[n]o una arma cum l'altra. E quigli maistri porterano corona in testa. Gli altri maistri mecionadi che serano drieto de lore porterano anche corona. E son chiamadi maistri remedy. E quigli che zugarano cum questi maistri e cum soy scolari sono chiamadi zugaduri. E gli scolari di questi maistri remedy porterano una diuisa sotto al genochio. E per capir le couerte e le prese segondo maistro remedio so. E sono tanti zoghi ch'el so maistro remedio poria fare in fino attanto che se troueza lo otrario del maistro remedio e di soy scolari. E questo otrario portera una corona in testa e una diuisa sotto el genochio. per che lo otrario del maistro e di scolari p[er] zo porta tanta diuisa lui ch'el maistro remedio cum tuti li soy scolari. ¶ In alguni loghi troueritte lo contrario subito drzedolo a remedio. E in algun loghi troueretu lo contrario drzedo di tuti li zoghi dello magistro remedio. Sapiando che qui lo contrario e fatto al maistro remedio, aquello otrario rompe tuti li soy zoghi de quella couerta ouero presa ch'ello fesse. Segondo che uoy troueritte dipinto e scritto sì bene che legiessimamente si poranno intender[e]. E començaremo prima a cauallo de lanza e de spada, e de abrazare. E poi drzedo de lanza ape. E poi de spada in arme. E poi de spada a doi man zogho largo. ¶ E poy stretto. E poi zogo de aza. E poi certi petiti. E poi de spada a una man. E poi zogho de abrazar ape. E poi zogho de daga. E per questo modo porite uedere tuta l'arte de armizar in questo libro che non se poza falare mente tanto drzano bene le glose sopra le figure dipinte.

☞ E questo magistro ha ligada una forte corda alla sella dello suo cavallo, çoè uno cavo e l'altro cavo si è ligado alo pe della sua lanza. Primo lo vol ferire e poy la lanza così ligada della parte stancha dello suo inimigo sopra la spalla la vole butar. ⁊ per ferlo ço dello cavallo strasinar.

2ᵛA ¶ *Aquesto magistro ha ligada una forte corda alla sella dello suo cavallo, zoe uno cavo, e laltro cavo si a ligado alo pe della sua lanza primo lo vol ferir e poij la lanza a cosi ligada della p<u>ar</u>te stancha delo suo inimigo sopra la spalla la vole butar, p<u>er</u> poterlo zo dello cavallo strasinare.*

This master has lashed a strong cord (that is, one end) to the saddle of his horse, and the other end is lashed to the foot of his lance. First he wants to strike, and then to put the tied part of the lance to the left of his enemy, throwing it over his shoulder, and thus to be able to pull him off his horse and onto the ground.

2ᵛB *No text.*

The Flower of Battle: MS M 383

3ʳA *No text.*

3ʳB *Io porto mia lanza in posta di dente di zenghiar perche io son bene armado, e si o curta lanza piu che lo compagno. ¶ E si fazo rasone de rebatere la sua lanza fora de strada, zoe ala taiversa[!] e non in erto. E si firiro Cum la mia lanza in la sua uno brazo in entro cum uno brazo dela mia asta. ¶ E la mia lanza discorrera in la sua persona. E lla sua lanza andera fora de strada lonze de mi e per tal modo faro como e dipento e scripto aqui.*

I carry my lance in the Position of the Wild Boar's Tusk because I'm well-armored and have a shorter lance than my companion. And so I make my strategy to beat his lance out of the way (this is, off to one side and not high), and thus I strike onto his lance with mine, with one arm's length of my haft one arm's length onto his. My lance will run into his person, and his lance will go out of the way far from me, and I'll do it in such a way as is written and drawn here.

3ʳC *No text.*

3ʳD *Aquesto si e lo contrario dello zogo de lanza che denanzi. Che qui uno corre contra laltro a ferri amoladi, e uno a curta lanza piu che laltro. ¶ Quando aquello che a curta lanza la porta la sua bassa in dente de zenghiar, Aquello che ala lanza longa debia similmente portarla bassa la sua, ¶ perche la curta non possa rebater la longa per lo modo che aqui dipento.*

This is the counter to the previous play of the lance: that here one runs against the other with sharp iron, and he has a shorter lance than the other. When he that has the short lance carries it low in Wild Boar's Tusk, he that has the long lance should similarly carry it low in the way that's drawn here, so that the short can't deflect the long.

Io porto mia lanza in posta di dente da zenghiar pche
io son bene armado e si o curta lanza piu che lo compa
gno. E si fazo rasone de rebatere la sua lanza fora
de strada zoe ala traversa e non p orto. E si firiro
cum la mia lanza in la sua uno brazo in ento cun
uno brazo de la mia asta. E la mia lanza discovera
in la sua psona. E la sua lanza andera fora de
strada lonze de mi e p tal modo faro como e dipento
e scripto aqui.

A questo sie lo contrario dello zogo de lanza che dnazi.
Che qui uno corre contra laltro a feri amoladi e
uno a curta lanza piu che laltro. Quando aqllo
che a curta lanza la porta la sua bassa in dente dzenghi
ar. A quello che a la lanza longa debia similmnte por
tar la bassa la sua. E pche la curta non possa rebater
la longa p lo modo che aqui dipento.

A questo e uno altro portar de lanza. A questo magistro
a cura la lanza e fia porta in posta de deura la sinistra
como noy vedite. p rebater e ferire lo compagno

S io lanzo mia lanza in lo peto dello tuo cauallo
lo tuo rebatere falo. E subito lanzada mia lanza.
la spada piglio p mia defesa. E cum tua lanza
non mi fazay offesa.

A cresa a questo magistro porta la sua lanza
in posta de coruo sinistra e p rebater la lanza
che lo compagno gli vole lanzare. E aquello
rebater chelo vol cum la lanza fare. A quello
cum uno bastone o cum spada far lo poria. Saluo
che sello buta sua lanza in lo peto de lo mio cauallo
lo mio rebater torneza fallo.

3ᵛA *No text.*

3ᵛB *Aquesto e uno altro portar de lanza. Aquesto magistro a curta lanza, e si la porta in posta de dona la sinistra como voij vedite, per rebater, e ferire lo compagno.*

This is another way to carry the lance. As you can see, this master has a short lance and carries it in the Position of the Noblewoman on the left, in order to deflect and then to strike his companion.

3ᵛC *Sio lanzo mia lanza in lo petto dello tuo cavallo, lo tuo rebatere fallo. E subito lanzada mia lanza la spada pigliro per mia defesa. E cum tua lanza non mi faraij offesa.*

If I throw my lance into your horse's chest, your deflection will fail. And as soon as I've thrown my lance, I'll take up the sword for my defense and you'll do me no offense with your lance.

3ᵛD *Anchora a questo magistro porta la sua lanza in posta di donna sinistra per rebater la lanza che lo compagno gli vole lanzare. E aquello rebatere che lo vol cum la lanza fare, Aquello cum uno bastone o curta spada far lo poria. Salvo che sello buta sua lanza in lo peto delo mio cavallo lo mio rebater tornera fallo.*

This master also carries his lance in the left Position of the Noblewoman in order to deflect the lance that his companion wants to throw. And that deflection which he wants to do with the lance, he could do with a stick or with a sword, too—except that if he throws his lance into my horse's chest, my deflection will turn to failure.

4ʳA No text.

4ʳB *Aquesto magistro che fuge non e armado, e si e bene a cavallo corrente, e sempre va butando le punte cum la sua lanza deriedo da si, per ferire lo compagno. E sello si voltasse della parte dritta ben poria intrar in dente de zenghiar cum sua lanza, overo in posta di donna la sinistra, e rebater, e ferire como si po fare, in lo primo & in lo terzo zogo de lanza.*

This fleeing master is unarmored and is on a running horse, and he is always throwing thrusts with his lance backward to strike his companion. And if he were to turn to the right side he could easily enter into the Wild Boar's Tusk with his lance (or into the Position of the Noblewoman on the left), and deflect and strike just as he can do in the first and third plays of the [mounted] lance.

4ʳC *Aquesto si e lo contrario dello zogho che denanci. Che questo magistro cum la lanza la porta bassa per ferire lo cavalo in la testa, o in lo petto, che lo compagno non po rebater cum la spada tanto basso.*

This is the counter to the previous play. And this master with the lance carries it low to strike the horse in the head or in the chest, because his companion can't reach so low[79] with his sword.

4ʳD *Aquesto portar de spada contra lanza e molto fino per rebatere la lanza, cavalcando dela parte dritta delo compagno. E aquesta guardia si e bona contra tute altre arme manuale, zoe contra azza, bastone spada, &c.*

This carry of the sword against the lance is very fine for deflecting the lance while riding to the right side of your companion. And this guard is good against all other hand-held weapons (that is, against the ax, the stick, the sword, etc.).

---

[79] Literally "beat/deflect so low".

A questo magistro che fuge no(n) e armado e sie bene a cavallo corrente e sempre va buttando le punte cu(m) la sua lanza deriedo da si per ferire lo compagno. E sello si voltasse della p(ar)te dritta ben poria intrar in dente de zenghiaro cum sua lanza. o vero i(n) posta di dona la sinistra o rebater e ferire como si po fare in lo p(r)imo zi(oe) lo terzo zogo de lanza.

A questo sie lo contrario dello zogho che denanza. Che questo magistro cum la lanza la porta bassa p(er) ferire lo cavalo in la testa o in lo petto che lo compagno non po rebater cum la spada tanto basso.

A questo portar di spada contra lanza e molto fino p(er) rebattere la lanza cavalcando dela p(ar)te dritta dello compagno. E questa guardia sie bona p(er) tute altre arme manuale zoe contra azza bastone spada.

E aquesto portar di spada e molto fino che ditto denanci
che se porta çi lanza como e ditto denanci. Che porto
la mia spada in posta dona finistra. E di questo mi
uene cum la lanza in resta p ferirmi e non el cauallo.
rebateo la sua lanza e cum mia spada lo feziro sença
fallo che la spada non po defendere basso p lo collo
del cauallo.

Anchora e aquesto uno altro contrario de lanza cot
spada. Che aquello de la lanza meti crcsta sua lanza
sotto lo suo brazo stancho poi che non sia rebattuda sua
lanza. E in tal modo porta ferir cum sua lanza a qllo
della spada.

4ᵛA *No text.*

4ᵛB *E aquesto portare di spada e molto fino che ditto denanci che se porta contra lanza come e ditto denanci. Che porto la mia spada in posta donna sinistra. E da questo mi vene cum la lanza in resta per ferirmi, e non el cavallo, rebatero la sua lanza, e cum mia spada lo feriro senza fallo che la spada non po defendere basso per lo collo del cavallo.*

This carry of the sword is very good (as was said earlier) and is carried against the lance (as was said earlier); I carry my sword in the left Noblewoman's Position. And if this one comes to me with lance in rest (to strike me rather than my horse), I'll deflect his lance and I'll strike him with my sword without fail. Note that the sword can't defend below the neck of a horse.

4ᵛC *No text.*

4ᵛD *Anchora e aquesto uno altro contrario de lanza contra spada. Che aquello dela lanza, meti e resta sua lanza sotto lo suo brazo stancho perche non sia rebatuda sua lanza. E per tal modo pora ferir cum sua lanza aquello della spada.*

Also, this is another counter of lance against sword. The lancer sets his lance in rest under his left arm so that his lance can't be deflected, and he can strike the swordsman with his lance like this.

5ʳA *No text.*

5ʳB *Aquesto cum la spada aspeta aquesto cum la lanza, e si lo aspeta in dente de zenghiare, como aquello cum la lanza gli vene apresso, lo magistro cum la spada rebati sua lanza in fora verso parte dritta. E acosi po far lo magistro cum la spada. chello po covirre[!] in uno voltar de spada.*

This swordsman awaits the lancer. He waits in the Wild Boar's Tusk as the lancer comes near, and then the master deflects his lance to the right side with his sword. And thus can the master do with the sword—that is, he can cover in a single turn of the sword.

5ʳC *Aquesto e lo contrario dello zogo de lanza, e spada che denanzi zoe che aquello cum la lanza fieri in la testa lo cavallo del suo inimigo zoe aquello dela spada perché non po rebater la lanza cum la spada si a basso.*

This is the counter to the previous play of the lance and sword: that is, that the lancer strikes at the head of his enemy's horse (that is, of the swordsman's), because he can't deflect a lance or sword which is so low.

5ʳD *No text.*

Aquesto cum la spada aspeta aquesto cum la lanza e si lo aspeta in dente de zenghiar como aquello cum la lanza gli vene apresso lo magistro cum la spada rebati sua lanza in fora verso parte dritta. E a cosi po farlo magistro cum la spada. Chello po covrire i uno volta de spada.

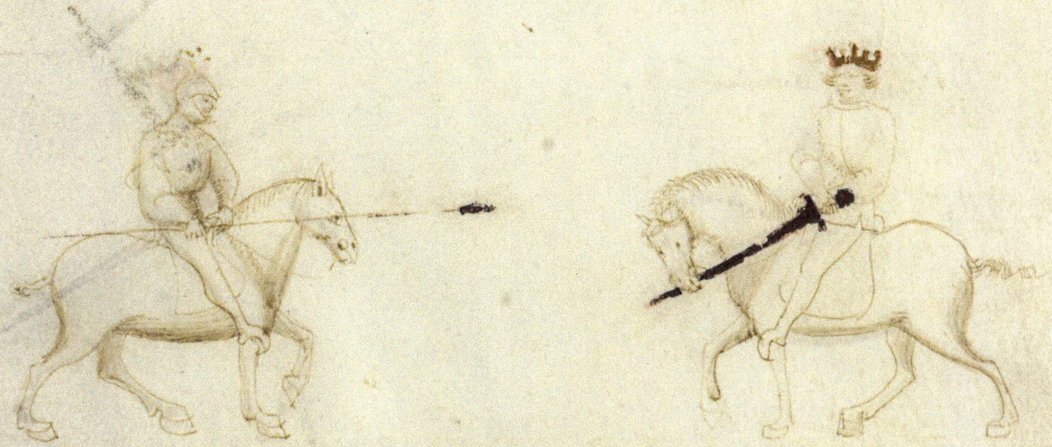

Aquesto e lo contrario dello zogo de lanza e spada che denanzi zoe che aquello cum la lanza fieri in la testa lo cavallo del suo inimigo zoe aquello dela spada perche non po rebatter la lanza cum la spada si abasso.

Aquesto portar de spada si chiama posta de coda longa e si e molto bona contra lanza e sp e contra ogni arma manuale. Cavalcando dela parte dritta de lo suo inimigo. Et vista bene amente che lo pute eli colpi riversi si dibano rebater in fora zoe ali traversa e non rexto. Elli colpi discendenti si dibano rebater anche in fora levando un pocho la spada dello suo inimigo e po fare gli zoghi segondo le figure depente.

Anchora aquesta propria guardia de coda longa si e bona quando uno gli vene in contra cum la spada a man riversa come vene questo mio inimigo e sapia che questa guardia e contra tuti colpi de pte dritta a di pte riversa e contra raschuno che sia drito o mancino. E qui de drieto comenzano gli zogi de coda longa che sempre rebuti p[er] lo modo che ditto denanzi in prima guardia de coda longa.

5ᵛA *No text.*

5ᵛB *Aquesto portar di spada si chiama posta de coda longa e si e molto bona, contra lanza e sp[ada], e contra ogni arma manuale. Cavalcando della parte dritta delo suo iminigo. E tente bene a mente che le punte e li colpi riversi si dibano rebater in fora zoe ala traversa e non in erto. E lli colpi di fendent si dibano rebater anche in fora levando un pocho la spada dello suo inimigo, e po fare gli zoghi segondo le figure depente*

This carry of the sword is called the Position of the Long Tail, and it's very good against lance and sword (and against all other handheld weapons) while riding to the right side of the enemy. Bear well in mind that thrusts and backhand blows should be beaten to the side (and not upward), and downward blows should also be beaten to the side (lifting the enemy's sword slightly); and [this guard] can make all the plays that are drawn in the following figures.

5ᵛC *No text.*

5ᵛD *Anchora aquesta propria guardia de coda longa si e bona quando uno gli vene in contra cum la spada a man riversa, come vene questo mio inimigo, e sapia che questa guardia e contra tuti colpi de parte dritta e di parte riversa, e contra zaschuno che sia drito o manzino. E aqui de driedo comenzano gli zoghi de coda longa che sempre rebati per lo modo che ditto denanci in prima guardia de coda longa.*

This same Position of the Long Tail is also good when someone comes against you with his sword on the left-hand side, as this enemy of mine does, and know that this guard goes against all forehand and backhand blows, and against anyone, be they right- or left-handed.

And hereafter commence the plays of the Long Tail, which always deflects in the way that was said earlier (in the first Guard of the Long Tail).

6ʳA *Aque⁸⁰ e lo primo zogho che esse dela guardia de coda longa che aqui denanci, zoe che llo magistro rebatti la spada delo suo inimigo, e metegli la punta in lo petto, o vole in lo volto come aqui dipento.*

This is the first play which belongs to the Guard of the Long Tail (which appeared earlier): that is, that the master deflects his enemy's sword and thrusts the point into his chest, or into his face (as drawn here).

6ʳB *Quisti doi magistri sono aqui incrosadi a tuta spada. E zo che po far uno po far laltro, zoe che po fare tuti zoghi de spada cum lo incrosar. Ma lo incrosar si e de tre rasone, zoe a tuta spada, e punta de spada. E chi e incrosado a tuta spada pocho gle po stare. E chi e incrosado a meza spada meno gle po stare. E chi a punta de spada niente gle po stare. Si che la spada si ha in si tre cose, zoe pocho, meno, e niente.*

Here these two masters are crossed at the fullness of the sword, and that which one can do, the other can do too—that is, he can do all the plays of the sword with this crossing.

But crossing is of three types (that is, from the fullness of the sword to the tip of the sword), and whoever is crossed at the fullness of the sword can withstand a little, and whoever is crossed at the middle of the sword can withstand less, and whoever [is crossed] at the tip of the sword can withstand nothing at all. So the sword, as such, has three parts—that is, a little, less, and nothing.

6ʳC *Aquesto si e lo segondo zogo che puo de quello rebater Io fiero a costui sopra la testa che vezo bene che ello non e armado in la testa.*

This is the second play which can deflect. I strike this man over the head, for I see well that he is unarmored on his head.

6ʳD *Aquesto e uno altro zogo che llo terzo che rebatuda la spada dello suo Inimigo ello la pigla cum la mane stancha e si gle feri la testa e acosi gle poria ferire de punta.*

Here is another play, which is the third that deflects the enemy's sword; he grasps with his left hand and strikes the [enemy's] head, and he could also strike with his point like this.

---

[80] *ue* is mostly effaced.

Aquesto zogho che esse dela guardia de coda longa che aqui denanci zoe che lo magistro rebati la spada delo suo inimigo e metegli la punta in lo petto o uole in lo uolto come aqui dipento.

Questi du magistri sono aqui incrosadi a tuta spada. E zoe po far uno po far l'altro zoe che po stare tuti zoghi de spada cum lo incrosar. Ma lo incrosar sie de tre rasone zoe a tuta spada, e a meza spada, e a punta de spada. E chi e incrosado a tuta spada pocho gli po stare. E chi incrosado a meza spada meno gli po stare. E chi a punta de spada niente gli po stare. Si che la spada si ha in si tre cose zoe pocho meno e niente.

Aquesto sie lo segondo zogho che puo de quello rebatazo fiero acostui sopra la testa che nezo bene che ello non e armado in la testa.

A questo e uno altro zogo che llo terzo che rebatuda la spada dello suo inimigo ello la piglia cum la mane stancha e si gli leua la testa e acosi gli poria ferire de punta.

Aquesto sie lo quarto zogo che lo scolar gli vole	Aquesto sie lo qnto che vol tor la spada al copa
fare zoe tore la spada p questo modo che vuy	gno. Cum lo mantenir dela spada l altro mante
possite vedere aqui depento.	nir laveza morto. delamane gli cadera la spada
	p certo.

Aquesto sie lo sexto zogo che fatta la conta cu	Aquesto sie lo septimo zogo che contrario del
lo rebater despada. Io gli buto lo brazo al collo	lo sexto e chi lo gli fa in la gamba a quello e
in lo voltar subito. Cum tutta la spada i tera lo	desso. Se lo compagno fosse armado non tefidera
buto senza dubito	in esso.
sie lo septimo zogo
lo et non ha logo.

6ᵛA *Aquesto si e lo quarto zogo che lo scolar gle vole fare zoe tore la spada per questo modo che vuij possite vedere aqui depento.*

This is the fourth play that the student wants to make—that is, take the sword in this way that you can see drawn here.

6ᵛB *Aquesto si e lo quinto che vol tore la spada al compagno. Cum lo mantenir dela spada, laltro mantenir lavera in erto. Dela mane gli cadera la spada per certo.*

This is the fifth play, in which he wants to take his companion's sword. With the hilt of his sword, he'll lift the other hilt up high. The sword will fall from [his companion's] hand for certain.

6ᵛC *Aquesto si e lo sexto zogo che fatta la coverta cum lo rebater de spada. Io glie buto lo brazo al collo in lo voltar subito. Cum tuta la spada in terra lo butero senza dubito. E lo mio contrario de dredo si e lo septimo zogo. Benche siando armado de farlo el non ha logo.*

This is the sixth play, which makes a cover with the deflection of the sword. I throw my arm around his neck and quickly turn, and I'll throw him to the ground, sword and all, without a doubt.

Hereafter is my counter, and it is the seventh play. Well that he hasn't achieved being armored.

6ᵛD *Aquesto si e lo septimo zogo che contrario dello ferire, chello gli fa in la gamba aquello e desso. Se lo compagno fosse armado non te fidar in esso.*[81]

This is the seventh play, which is the counter—that is, the strike that he makes at the other one's leg. If your companion were armored, you couldn't rely on this.

---

[81] This paragraph is partially effaced and hard to read.

7ʳA *Aquesto si e lo ottavo zogo che contrario de tuti li zoghi che mi sono denanci. E maximamente deli zoghi de spada a cavallo, e deli lor magistri che sono in guardia de coda longa. Che quan li magistri, o scolari stano in la ditta guardia e io tra una punta, o altro colpo. E subito elli me rebateno, o taio, o punta che faza. Quando elli me rebateno subito io do volta ala mia spada e cum lo pomo mio io fero in lo volto. E poij passo cum la mia coverta presta. E cum lo riverso tondo gli fero dredo la testa.*

This is the eighth play, and it is the counter to all the previous plays, and especially to the plays of the sword on horseback and to the masters that are in the Guard of the Long Tail.

When the masters or students are in said guard and I strike with a thrust or another blow, they immediately deflect it (whichever cut or thrust I throw). When they deflect it, I quickly give a turn to my sword and strike them in the face with my pommel. And during my cover, I can pass through quickly and strike them behind the head with a backhand middle cut.

7ʳB *Lo nono sone che fazo lo contra lo contrario che me denanci. Che quando ello da volta ala sua spada, subito lo mio mantenere meto como vuij vedeto depento, che cum lo pomo in lo volto non me po ferire, e sio levo la spada in erto e dello riverso io piglo volta. Ben poria essere che la spada ti sara tolta. E si aquello mi falla che io non lo faza, dello riverso dela spada ti daro in la faza overo dello pomo ti firiro in la testa tanto faro una volta presta. Aqui finisse lo zogo a cavalo de spada a spada chi piu ne sa men dia una bona derada.*

I'm the ninth, who does the counter to the previous counter so that when he gives a turn to his sword, immediately I thrust my hilt (as you see drawn here) so that he can't strike me in the face with his pommel. And if I raise my sword high and give a turn to the left, it could very well be that his sword will be taken from him. And if that fails me and I can't do it, I'll make such a quick turn that I'll give to his face with the back edge of my sword (or I'll strike him in the head with my pommel).

This finishes the mounted play of sword against sword, and whoever keeps it in mind will give a good deal.

A questo sie lo ottavo zogo che contrario de tuti li zoghi che mi sono denanci. E maximamente deli zoghi de spada a cavallo e deli soi magistri che sono in guardia de coda longa. Che qui li magistri o scolari stano in la dita guardia e io tra una punta o altro colpo. E subito elli me rebateno / o taio o punta che faza. Quando elli me rebateno / subito io do volta a la mia spada e cu lo pomo mio io fiero in lo volto. E poy passo cu la mia coverta pista. E cum lo riverso tondo gli fiero dredo la testa.

Lo nono sone che fazo lo contra lo contrario che me denega. Che quando allo dar volta a la sua spada subito lo mia manteniri meto como vuy vedeto depento che cu lo pomo in lo volto non me po ferire e si lovo la spada in cito de allo riverso io piglo volta. Convera essere che la spada ti sara tolta. E si aquello m'e falla che io no lo faza. Dello riverso dela spada ti daro in la faza o vero dallo pomo ti fiero in la testa tanto fiero ma volta pista a qui fosse lo zogo a cavallo de spada a spada chi piu ne sa men dia una bona derada.

Aquesto e zogo de abrazar e tanto e a dire zogo de abrazar che zogo d'brazar e si fa p tal modo. Quando
uno te fuzi e dela pte stancha tu gli ven apresso
Cum la man dritta tu lo pigli i le sguanze dello bacinetto e sollo e desarmado p gli cavili. Ouero p
lo brazo dritto p dedo le sue spalle p tal modo lo
farai atrauestir. Che interra lo fazay andare.

Aquesto e contrario dello zogo che dinanci me ua
p tal modo aquesto contrario cum tal presa se fa
zo e che subito quando ello p dedo lo piglia la man
dalla briglia debia subito scambiare. E cum lo brazo
stancho p tal modo lo di piglare.

Aquesto scolar uole buttar questo da cauallo zoe che
lo piglia p la staffa e cauallo incerto. Sello no ua
interra macro stara p certo. Saluo che se no e a
cauallo ligado. Aquesto zogo no po esser fallado
solo non lu lo pe in la staffa p lo collo del pe lo pigli
che pui vale leuando p erto come denanci e dito, fate aquello che denanzi aqui scrito.

lo contrario aqui del zogo che denanzi aparechiado che
se uno te piglia p la staffa ouero p lo pe. Butagle
lo brazo allo collo. aquesto subito far sede. E p
tal modo lo porai de smuntare da cauallo. Ben fai
a questo ello andera p terra senza fallo.

7ᵛA *Aquesto e zogo de abrazar e tanto, e a dire zogo de abrazar che zogo di braci, e si fa per tal modo. Quando uno te fugi e dela parte stancha tu gli ven apresso Cum la man dritta tu lo pigli in lo sguança dello bacinetto, e sello e desarmado per gli cavili, Overo per lo brazo dritto, per dredo le sue spalle, per tal modo lo faraij arivesare, Che in terra lo faraij andare.*

This is a play of grappling, and inasmuch as it's a play of grappling, it's a play of the arms. It's done like this: when someone flees from you and you come up behind him from the left side, grab him on the cheek of his helmet with your right hand (or, if he is unhelmed, grab him by the hair or by the right arm from behind his shoulder), and in this way you'll make him fall backward such that you'll make him go to the ground.

7ᵛB *Aquesto e contrario dello zogo che dinanci me va per tal modo aquesto contrario cum tal presa se fa zoe che subito quando ello per dredo lo piglia la man della brigla debia subito scambiare. E cum lo brazo stancho, per tal modo lo de piglare.*

This is the counter to the previous play. This counter goes with the hold that was done like this: that is, that quickly when he grabs him from behind, [the master] should immediately exchange the hand on the reins, and he should grab him with his left arm like this.

7ᵛC *Aquesto scolar vole butar questo da cavallo zoe che lo pigla per la staffa e levalo in erto. Sello non va in terra in aere stara per certo. Salvo che se non e a cavallo ligado. Aquesto zogo non po essere falado. Selo non ha lo pe in la staffa per lo collo del pe lo pigla che piu vale levando in erto come denanci e ditto, fate aquello che denanzi aqui scrito.*

This student wants to throw this one from his horse—that is, he grabs him by the stirrup and lifts him up. If he doesn't go to the ground, he would clearly be floating in the air! Assuming he isn't lashed to his horse, this play can't fail.

If he doesn't have his foot in a stirrup, grab him by the ankle and it'll be even easier to lift him up than I said earlier, so do as was written here earlier.

7ᵛD *Lo contrario aqui del zogo che denanci aparechiado et se uno te pigla per la staffa overo per lo pe, Butagle lo brazo allo collo, aquesto subito far se de. E per tal modo lo porai descavalcare da cavallo. Stu fai aquesto ello andera per terra senza fallo.*

This here is the counter to the previous play: if someone grabs you by the stirrup (or by the foot), throw your arm around his neck. You should do this quickly, and you could dismount him from his horse like this—if you do this, he'll hit the ground without fail.

8ʳA *Aquesto e uno zogo de butare uno in terra cum tuto lo cavallo zoe che lo magistro cavalcha dela parte dritta dello suo Inimigo, e buta lo suo brazo dritto per sopra lo collo dello suo cavallo. E pigla la brena delo so cacavallo[!] apresso lo morso, revoltando la testa dello cavallo in erto e llo suo debia speronare che lo suo cavallo cum lo suo petto fiera in gropa overo in gli fianchi del suo cavallo. E per tal modo cadera cum tuto⁸² lo cavallo. Lo contrario de questo magistro che vole butare in terra lo suo inimigo cum tuto lo cavallo. Si e aquesto che subito quando lo magistro pigla la sua brena. Che ello debia butare lo brazo al collo per modo che fa lo quarto zugadore che me denanci per tal modo andera in terra.*

This is a play of throwing someone to the ground, horse and all: that is, the master rides to the right side of his enemy, throws his right arm over the neck of his [enemy's] horse, and grabs the bridle of his horse behind the bit; he rotates the horse's head upward, and he should spur his horse (with his foot striking its rump or flanks), and in this way [his enemy] will fall, horse and all.

The counter to this master that wants to throw his enemy to the ground, horse and all, is this: that immediately when the master grabs [my] bridle, [I] should throw my arm around his neck in the way that the fourth player before me does, and in this way, he'll go to the ground.

8ʳB *Aquesto e uno zogho de tore la brena dello cavallo de la maane⁸³ delo compagno per modo che voij vedite aqui dipento, lo scolar quando ello se scontra cum uno altro da cavallo, ello gle cavalcha dela parte dritta, e butagli lo suo brazo dritto per sopra lo collo del cavallo, e pigla la sua brena⁸⁴ apresso la sua man sinistra cum la sua man riversa. E tra la brena delo cavallo dela testa. E aquesto zogo e piu seguro armado che disarmado.*

This is a play of taking the reins of the horse from your companion's hand in the way that you see drawn here. The student, when he goes against another on horseback, rides to the right side and throws his right arm over the horse's neck, grabbing its reins near [the enemy's] left hand with his hand reversed, and so pull the bridle off the horse's head. And this play is more certain in armor than unarmored.

8ʳC *Aqui sono tri compagni che volemo ferire aquesto magistro lo primo vole ferire sotto man che porta sua lanza a meza lanza laltro porta sua lanza a restada a tuta lanza, lo terzo lo vole alanza cum sua lanza E si a de patto che nesuno non debia fare piu de uno colpo per homo, anchora debano fare a uno a uno.*

Here are three companions who want to strike this master: the first wants to strike underhand and so carries his weapon at the middle of the lance; the other carries his weapon in rest at the fullness of the lance; the third wants to throw his lance. And they are agreed that none should strike more than one blow per man, and that they should do it one by one.

8ʳD *Vegna a uno a uno chi vole venire, che per nesun di qui non per partir. Anche in dente de zenghiar sono posto per aspetare, quando la lanza contra me vignira portada overo de man zitada, subito io schivo la strada zoe che io acresco lo pe dritto fuora de strada, e cum lo stancho passo ala traversa rebatendo la lanza che mi vene per ferir. Si che de mille una non poria falir. Aquesto io fazo cum la chiavarina cum bastone cum spada lo faria. E la defesa chio Fazo le contra le lanze contra spada contra bastone aquello faria li mie zoghi sono di dredo, guardagli chio ven prego.*

Come one by one, whoever wants to come, because I'll run from no one. I'm also standing in Wild Boar's Tusk to wait, and when the lance will come against me (carried or thrown by hand) I quickly dodge its path—that is, I advance my right foot out of the way and I step to the side with my left, deflecting the lance that comes to strike me such that out of a thousand, I couldn't fail even once. I do this with the bearded javelin; I would do it with the staff or with the sword, too, and the defense that I make against the spear, I would also make against the sword or against the staff. My plays are hereafter, I ask you to view them.

---

[82] Corrected from *i*; probably intended to be a *u* but looks like an *a*.
[83] Overwritten and difficult to decipher.
[84] Written over a previously-effaced word that can't be deciphered.

Aquesto e uno zogho de buttare uno in terra cum tuto lo cauallo zoe chello magistro caualcha de la p[ar]te dritta dello suo Inimigo e butta li lo suo brazo dritto p[er] sopra lo collo dello suo cauallo. E piglia la brena dello so ca- uallo apresso lo morso, reuoltando la testa del caual- lo i[n] erto e llo suo debia sperzonare chello suo cauallo cum lo suo petto fieza i[n] la gropa o[uer] i[n] gli fianchi del suo cauallo. E p[er] tal mo[do] andera cu[m] tuto lo cauallo i[n] terra. Lo contra[r]io de questo magistro ch[e] uole buttare i[n] terra lo suo Inimigo cum tuto lo cauallo. Fie aquesto ch[e] subito quando lo magistro piglia la sua brena. Ch[e] ello debia buttare lo brazo al collo p[er] modo che fa lo q[ua]rto zugadore che me denaci. E p[er] tal mo[do] andera i[n] terra.

Aquesto e uno zogho de torre la brena dello cauallo de la rauene delo compagno p[er] modo che uoy uedete aqui di p[re]sent[e] lo scolar qua[n]do ello se sco[n]tra cu[m] uno altro da cauallo, ello gle caualcha de la p[ar]te dritta e butta gle lo suo brazo dritto p[er] sop[ra] lo collo del cauallo. E piglia la sua brena apresso la sua ma[n] sinistra cu[m] la sua ma[n] rouersa. E tra la brena delo cauallo dela testa. E aquesto zogo e piu seguro armado che disarmado.

Aqui sono tri compagni che uolemo ferire aquesto magistro lo primo uole ferire sotto man p[or]ta sua lanza a meza lanza l'altro porta sua Lanza aresta a tuta Lanza. Lo terzo lo uole a lanza cu[m] sua lanza E sic de patto che nessuno no[n] debia fare piu de uno col- po p[er] homo. anchora debi fare ad uno ad uno.

Degna cosa e dono chi uole uenire che nessu[n] di lor no[n] p[er] p[ar]e. Anche inte[n]de de zeng[h]iar so[n] posto p[er] scolar qui la lanza çita me uegnira portada ouer de mezo zitada subito io schiuo la strada zoe ch[e] io acresto lo pe dritto fuora de strada e cu[m] lo sta[n]co p[asso alatrauerso ro- batendo la laza che mi uene p[er] ferir. Siche de mille una no[n] pora fallir. Aquesto io farò cu[m] la chiauarina cu[m] bastone cu[m] spada lo farò. E la defessa ch[e] io fazzo le g[r]a[n] le lanze cont[r]a spada o[uer] bastone a q[uello] ferire li mie zoghi sono di dredo. Guardali ch[e] no[n] p[er]o.

Aquesto zogo sie de lo magistro che denanci che aspetta
cum la chiavarina questi da cavallo in dente de zenghiar. In peter foza de strada e rebater chello su-
ello intra in questo zogo. e p chello sia intese, io lo
fazo in suo logo. Che cum taglo e punta lo posso ferir in la testa. Tanto porto la mia chiavarina bon presta.

Anchora e aquesto zogo dello dito magistro che denanci imposta de zenghiar in suo scambio io fazo
aquesto chello po fare, quando la lanza e rebatuda
Io volto mia lanza e si lo fiero cum lo pedale. E
che aquesto fazo sie tempado ede tuto azalo

8ᵛA *Aquesto zogo si e de lo magistro che denançi che aspetta cum la chiavarina quisti da cavallo in dente de zenghiar. In pasar fora de strada e rebater chello fa ello intra in questo zogo, e p<u>er</u>chello sia inteso, io lo fazo i<u>n</u> suo logo. Che cum taglio e punta lo posso ferir[85] i<u>n</u> la testa. Tanto porto la mia chiavarina ben presta.*

This play belongs to the previous master, who awaits the horseman with his bearded javelin in Wild Boar's Tusk. In stepping out of the way and deflecting as he does, he enters into this play; so that it's understood, I do according to his word. I could strike him in the head with the edge and the point, since I carry my javelin with such readiness.

8ᵛC *Anchora e aquesto zogo dello ditto magistro che denanci in posta de zenghiar in suo scambio io fazo aquesto chello po fare, quando la lanza e rebatuda Io volto mia lanza e si lo fiero cum lo pedale. Che aquesto ferro si e temp<u>er</u>ado e de tuto azale.*

This play is also from the foresaid master who was previously in the Position of the Wild Boar. I do this in his place (which he could've done too): when the lance is deflected, I turn my lance and I strike him with the butt (since it's made of tempered steel).

---

[85] *e* changed to *r*.

The Flower of Battle: MS M 383

9ʳ *Aqui comenza larte de nobele arma chiamada lanza principio de bataglia a cavallo, e a pe e sua usanza. E chi la guarda cum so bello penone e polito de grande paura doventa smarido. E la fa grande punte, e pericolose forte. E cum una sola po dar la morte. E si lo primo colpo el suo debito ella fara, Azza spada e daga de impazo tute le cavara.*

Here begins the art of the noble weapon called Lance; its use is in the beginning of duels, on horse and on foot. And whoever sees it with its dashing pennant should be frightened with great dread. And it makes great thrusts which are dangerously strong, and it can give death with a single blow. And if in the first blow it makes its due, then ax, sword, and dagger will all be upset.

9ʳA *Nuij semo tri magistri in guardia cum nostre lanze e convegnemo piglare aquelle della spada. E io non son lo primo che in tuta porta di ferro son posto per rebater la lanza dello zugadore tosto zoe. Che passaro cum lo pe drito ala traversa fora de strada. E atraversando la sua lanza rebatero in parte stancha. Si che llo passar e llo rebater se fa in uno passo cum lo ferire. Aquesta cosa che non se po falir.*

We are three masters in guard with our lances, and it suits us to hold those of the sword.

And I'm not[86] the first, and I'm set in the Full Iron Gate to quickly deflect the player's lance—that is, that I'll step out of the way to the side with my right foot, and crossing his lance, I'll beat it to the left side such that the step and the deflection are done in the same movement as the strike. This is a thing that can't fail.

9ʳB *In meza porta di ferro io me ho posto cum curta lanza lo rebater e llo ferire, e sempre mia usanza. E vegna chi vole cum longa lanza, o stanga, che rebater cum passo lo ferire non mi mancha. Che tute le guardie che stano fora de strada, cum curta lanza, e curta spada, Sono soficienti aspetar, ogni arma manuale longa. E aquelle della parte drita, covrano, e cum coverta passano, e meteno punta. E lle guardie de parte sinistra covrano, o rebateno, e di colpi fierano e non po metere acosi ben punta.*

I've set myself in the Middle Iron Gate with short lance, and my method is always to deflect and to strike. Come whoever wishes with long lance or staff, and I'll deflect with a step and I won't botch the strike. And all the guards that stay out of the way with short lance or short sword are sufficient to await any long hand-held weapon. Those of the right-side cover, and during the cover they step and thrust the point, while the guards of the left side cover or deflect and then strike with blows (and can't thrust very well with the point).

9ʳC *Io son in la nobele posta de fenestra dextra che in rebater, e ferir sempre io son presta. E de lanza longa me curo pocho. Ancora cum la spada e aspetaria la longa lanza siando in questa guarda che ogni punta rebate e si la intarda. E llo scambiar de punta io posso fare. E llo rebater a terra non se po falar. In lo zogo che ne dredo volemo finire.*

9ʳD *In questo zogo finisseno li tre guardie che denanci son zoe, tuta porta di ferro, e porta di ferro la mezana, e posta di fenestra dextra la soprana in questo zogo elle finisseno li zoghi e la lore[!] arte. Como io fiero a costui in lo volto[87] per lor parte. Aquesto e lo contrario deli tri magistri de la lanza che finisseno in questo zogo, el modo voglo dire. Quando gli magistrij credeno la mia lanza fora de lor persona cazare io do volta ala mia lanza, e ferischo cum lo pedale. ~~che a~~ E acosi o bon Ferro in lo pedale chala punta, gli zoghi de quisti magistri pocho me monta.*

I'm the noble right Position of the Window; I'm always quick to deflect and to strike, and I care little about long lances.

I would also stand in this guard with a sword to await a long lance, so that every thrust would be deflected and retarded. I could make an exchange of the thrust or beat them to the ground, too (which can't fail). We want to finish in the play hereafter.

The previous three guards (that is, the Full Iron Gate, the Middle Iron Gate, and the high right Position of the Window) should finish in this play. In this play they finished their plays and their art, and I strike this man in the face on their behalf.

This is the counter to the three masters of the lance that finish in this play, and I want to tell you how: when the masters believe my lance to be pushed away from their persons, I give a turn to my lance and strike with its butt, since the iron in the butt is as good as in the point. The plays of these masters bother me little.

---

[86] Unclear why *non* is here; it's not present in the Getty and seems contrary to the intent of the statement.
[87] Word partially effaced.

Aqui comenza l'arte de nobele arma chiamada lanza, principio de bataglia a cavallo, e a pe e sua usanza. E in la guarda cum so bello penone e polito de grande paura douenta smarido. Ella fa grande punte e pericolose forte, cum una sola po dir la morte. E si lo primo colpo el suo debito ella fae. Azza spada e daga de imparo tute le cauara.

Nuy semo tri magistri in guarda cum nostre lanze, e come gnemo pigliar aquelle della spada. E io non son lo primo che in tuta porta di ferro son posto, per rebater la lanza dello zugadore tosto zoe. Che passaro cum lo pe drito ala trauersa fora de strada. E atrauersando la sua lanza rebatero in parte stancha. Si chello passar ello rebater se fa in uno passo cum lo ferir. A questa cosa che non se po falar.

In meza porta di ferro io me ho posto cum curta lanza lo rebater e lo ferire e sempre mia usanza. E uegna chi uole cum longa lanza o stanga che rebater cum passo lo ferire non mi mancha. Che tute le guardie che stano fora de strada, cum curta lanza e curta spada sono soficienti a spetar ogni arma manuale longa. E aquelle della parte dritta courano e cuz couita passano e meteno punta. Ello guardie de parte sinistra courano o rebateno e di colpi ficano end per moteue acosi ben punta.

Io son in la nobele posta di fenestra dextra che in rebater e ferir sempre io son pista. E de lanza longa me curo pocho. Ancora cum la spada e aspetaria la longa lanza stando in questa guarda che ogni punta rebate e sila in tuda. Ello scambiar de punta io posso fare. Ello rebatter a terra non se po falar. In lo zogo che ne dredo volemo finire.

In questo zogo finisseno li tre guardie che denanzi son zoe tuta porta di ferro e porta di ferro la mezana e posta di fenestra dextra la soprana in questo zogo elle finisseno li zoghi e la loro arte. Como io fazo acostui in lo usito per parte. A questo e lo contrario delli tri magistri della lanza che finisseno in questo zogo el modo uoglio dire. Quando gli magistri credeno la mia lanza fora de loro persona cazere io do volta alla mia lanza e feriscolo cum lo pedale. Che a cosi o bon ferro in lo pedale che ala punta. Gli zoghi de questi magistri pocho me monta.

Nuy semo tre guardie di parte riuersa e io son la prima in denti de zenghiaro. Aquelle guardie che sone da parte dritta fano aquello che fazemo de la riuersa. Nui passamo fora de strada e in acresendo lo pe che denanci como ditto fora de strada. E de nostre punte de parte riuersa fazemo dezada. E tute de parte dritta e riuersa conuegnemo in punta rebatendo finir. Che altra offesa cum la lanza no ne po seguir.

Io in posta de nera croce io aspeto tume tropo apresso zoga netto. Io pe dritto che me denci in dredo lo tornero. E la tua lanza rebatero fora de strada in uerso man dritta. La mia punta no falira la tua sera falita

In posta fenestra sinistra io son apareclado se non te fiezo cum punta tu hay cum bon mercato. La punta tegnero erta colli brazzi bassi portero cum lo pe che dredo cum quello io passaro fora de strada a man riuersa. La punta ti metero in lo volto senza nesuna defesa. El zogo che me dredo noy tre magistri aquello possemo noy fare una volta lo piu no lo toray piu proue.

Lo zogo de la lanza aqui finisse che io lo fiero de le punte riuersa de li zoghi me impazo. A queste tre guardie che sono denanci fano pensar de lanza longa o curta donde la falsir che le sono guardie de si grande difesa che uno uergo le fano de fesa e ofesa. E lo otrario de questa punta e quello bene se po fare. Quando eu lo ferir la punti se rompe lo pe de lo se uoltar. De lo zogo de la lanza basta po a questo bastare.

9ᵛA *Nuij semo tre guardie di parte riversa, e io son la prima in dente de zenghiar. Aquelle guardie che sone de parte drita fanno, aquello che fazemo della riversa. Nuij passamo fora de strada innanci acresendo lo pe che denanci como ditto, Fora de strada. E de nostre punte de parte riversa fazemo derada. E tute de parte dritta, e riversa convegnemo in punta rebatendo finir, Che altrᵃ offesa cum la lanza non ne po seguir.*

We are three guards of the left side and I'm the first, in the Wild Boar's Tusk. We do the same on the left side as those guards that are on the right: we step out of the way, first advancing the rear foot (also out of the way, as was said), and with our thrusts on the right side we strike a bargain. And to finish the deflection, both the right side and the left must thrust so that other offense with the lance can't follow.

9ᵛB *Io in posta de vera croxe, io aspeto, tu me tropo apresso zoga netto. Lo pe deritto che me denanci in dredo lo tornero. E la tua lanza rebatero fora de strada, in verso man dritta. La mia punta non falira la tua sera falita*

I wait in the Position of the True Cross. You're too close for my play, so I step backward with my right foot (which is in front), and I'll beat your lance out of the way to the right-hand side. My thrust won't fail, but yours will.

9ᵛC *In posta fenestra sinistra io son aparechiado se non ti fiero cum punta tu naij ł bon mercato. La punta tegnero erta, e lli brazi bassi portero cum lo pe che dredo cum quello io passaro fora de strada a man riversa. La punta ti metero in lo volto senza nesuna defesa. El zogo che me dredo noij tre magistri aquello possemo noi fare una volta lo provi non lo voraij piu provare.*

I appear in the left Position of the Window; if I don't strike you with my point, you'll have a good deal. I'll hold the point high and I'll carry my arms low, and I'll step out of the way to the left-hand side with my rear foot. Then I'll thrust my point in your face without any possible defense.

We three Masters can all make the next play; once you taste it, you won't want to try any more.

9ᵛD *Lo zogo dela lanza aqui finisse che io lo fazo della parte riversa de lor zoghi me impazo. Aqueste tre guardie che sono denanci fano pensir, de lanza longa, o curta, de non la falir, che le sono guardie de si grande difesa, che in uno vargo le fanno defesa e ofesa. E llo contrario de questa punta, aquello bene si po fare. Quando cum quello ferir la punta se rompe lo pedale se de voltar. Dello zogo dela lanza bene po aquesto bastare.*

The play of the lance finishes here, and I make it from the left side; I'm besotted with their plays. These three previous guards make me think that they won't fail against long or short lance, for they are guards of great protection (in that in one jaunt they make both defense and offense).

And the counter to this thrust can be done well: when the thrust is broken with that strike, the butt should be given a turn [forward]. This may be enough of the play of the lance.

The Flower of Battle: MS M 383

10ʳ *Aqui comenza la spada de armizare. Ben sera magistro chi tali zoghi sara fare. Gli magistri sono sie e zaschuno in guarda. De covrir e ferir non farano niente tarda. E chi piu savera i̱n questa lor arte, de tuti lor zoghi che segueno avera parte.*

Here begins the sword in armor. Great will be the master who can make these plays. These masters are six and each one's in guard; they won't delay in covering and striking. And whoever knows most in this, their art, will have a part in all of the following plays.

10ʳA *In posta breve la se̱rpentina io voio venire. Se tu non e bene armato ben te lo faro sentire pe̱r ferir de punta meior delle altre guardie mi tegno. Pe̱rche delli tagli cum la crose mi segno e niente mi pon fare. In arme e senza arme lo voio provare.*

I want to come in the Short Position, the Serpent. If you aren't well-armored, I'll make you feel it! I hold myself better than any other guard for striking with the point. Because of my edges, I sign myself with the cross, and you can do nothing to me. I want to prove it in armor and without.

10ʳB *Io son posta de vera crose che contra ti voio fare. In mi la tue punte no̱n pon entrar. De ti me coviro in lo passare che faro. E de punta te feriro senza falo. Che ti e lle altre guardie pocho me pono fare. Tanto so bene lo armizare. Che non posso falire lo incrosare. Che i̱n lo passare e i̱n lo incrosar, e i̱n lo ferire, larte vole a questo non falir. E ro̱mpo tute tue punte, e no̱n faliro pe̱r certo vene oltre, e tra voij basso, la punta voi erto.*

I'm the Position of the True Cross, which I want to make against you. Your thrusts won't enter into me: I cover myself from you during the step that I take, and my point will injure you without fail, so that you and the other guards can do little to me. I know armored fencing so well that I can never fail in the crossing, since without fail the art wants these things: stepping, crossing, and striking. And I break all your thrusts and I certainly won't fail: I come over and through, I go under the point and upward.

10ʳC *Sompuo[!] serpentino son lo soprano, e ben armato grande punte butto subito sotto mane, che son i̱n erto e torno al piano. Una forte punta te butiro cu̱m la passare. Ella e mia arte che lo so ben fare, delli tuoi tagli no̱n me curo niente tanto so in larte. Che de grande punte io te daro grande parte.*

The Supreme[88] Serpent, I'm high and well-armored. I quickly make great underhand thrusts because I'm high and return below. I'll drive a strong thrust at you with a step: this is my art and I know how to do it well. I don't care about your edges, for I know the art and I'll give you the better part of my point.

10ʳD *De ferro son chiamada mezana porta pe̱rche i̱n arme e senza e fazo le punte forte. E passaro fuora de strada cum lo pe stancho, e ti mezo una pu̱nta i̱n lo volto. Overo che la punta e cum lo taglo enfra gli toij braçi intrero pe̱r modo che io te metiro i̱n la ligadura mezana. In quella che dena̱nci dipenta e nominada.*

I'm called the Middle Iron Gate because I give strong thrusts in armor or without. And I'll step out of the way with my left foot and thrust my point in your face, or I'll enter [with] my point and with the edge between your arms and force you into the Middle Bind (which was previously named and drawn).

---

[88] *Sompuo* isn't a word as far as I can tell, and doesn't really even resemble one. The Getty spells it *Sompno*, which is also not a real word but has been interpreted as *sonno* (meaning "sleep") based on the rubric. Because *sonno* doesn't make much sense, I'm reading it as *sommo* here, but that's a wild guess.

Aqui comenza la spada de aznizare. Ben seza magistro chi tali zoghi sara fare. Gli magistri sono sie e zaschuno in guarda de contro e ferir non farano niente tarda. Sehi piu sauera i questa l'arte. de tuti lor zoghi che seguen aueza parte.

Imposta breue la serpentina io uoio deuare. Se tu non e bene armato bentelo faro sentire. p fezir de punta meior delle altre guadie mi tegno. Che delli tagli cum la croxe niu segno e niente mi pon fare. In arme e senza arme lo uoio prouare.

Jo son posta de uera croxe che contati uoio fare. In mi le tue punte no pon intrare. de ti me couriro in lo passare che fazo. E de punta te ferirò senza fallo. Che ti elle altre guardie pocho me pono fare. E tanto so bene lo aznizare. Che i non posso fallir lo inzosare. Che ì lo passare e ì lo inzoso, e ì lo ferire. L'arte uole aquesto non fallir. E zopo tute tue punte, e no fallo p zerto bene oltra, e tra non basso la punta uoi zerto.

Compuo serpentino son lo sopratuo, e ben armado grande punte butto subito sotto mane, che son p zerto eterno alpino. Una forte punta te butaro cû la passare. Ella e mia arte che lo so ben fare. delli tioi tagli no me curo niente tanto so in arte. Che de grande punte io te daro grande parte.

De ferro son chiamada mezana porta poke l'arme e senza, e fazo le punte forte. E passazo fuora de strada cum lo pe stanch, o ti mezo una punta ì lo uolto. Ouero che la punta e cû lo taglio ensa gli toi braci intrero p modo che io te tiro ì la lingadura mezana. In quella che denâzi dipenta e nominada.

Posta sagittaria son p nome chiamata. grande punta io zeto passando fora de strada. E si me contra colpo o taglio io fazo bona coverta e subito io fiero lo mio contrario. A questa e' mia arte che non fallo.

di meza croxe son la strada posta. Zoche alla posta voluntiera lo fazo. Done coverte e punte e tagli fazo p usanza. Sempre schivando gli colpi fora de strada. E de mie colpi io fazo grandissima derada.

De posta de meza croxe io son ensido cü questa coverta passando fora de strada ala traversa. E di questa coverta si vedra quello a quello che io posso fare. P gli mie scolari lo posso mostrare. Che li faranno li mie zoghi i coplimito. A qui gli che son da combater ha oltraza. La te monstraramo senza dubitanza.

E son lo p'mo scolar delo magistro che me donato e aquesta punta fazo che la ose della sua coverta. Anchora digo che della posta de meza croxe ede posta de croxe bistrada posso fare aquesta punta e digo de subito zoe come lo zugadore tra una punta alo magistro ouero scolar che fosse i le due guardie ouero poste lo magistro ouero scolar de andare basso cü la psona e passar fora de strada atrauersando la strada ecu la punta cru al volto ouero al petto ecu lo mantenir della spada abasso como dipinto aqui.

10ᵛA *Posta sagittaria son per nome chiamata, grande punte io zeto passando fora de strada, e si me ven contra colpo o taglio io fazo bona coverta, e subito io fiero lo mio contrario. Aquesta e mia arte che non svario.*

The Archer's Position, by this name I'm called. I give great thrusts while stepping out of the way, and if you come against me with a blow of your edge, I make a good cover and I quickly strike my counter. This is my art and it doesn't change.

10ᵛB *Di vera crose son bastarda posta. Zo che ella po far voluntiera lo fazo. Bone coverte e punte e tagli fazo per usanza. Sempre schivando gli colpi fora de strada. E deli mie colpi io fazo grandissima derada.*

Of the True Cross, I'm the Bastard Position; that which she can do, I choose to do too. For my strategy, I make good covers, thrusts, and cuts, always while dodging out of the way of blows, and with my blows I strike my greatest bargain.

10ᵛC *De posta de vera crose io son ensudo cum questa coverta passando fora de strada ala traversa. E di questa coverta si vedera quello aquello chio posso fare. Per gli mie scolari lo posso mostrare. Che li fazano li miei zoghi in complimento. Aqui gli che son da combater ha oltranza, larte⁸⁹ monstrarano senza dubitanza.*

I come from the Position of the True Cross with this cover, stepping out of the way to the side. And you'll see what I can do from this cover: I can show it through my students, because they make my complementary plays. Here, those that are to fight in the extreme will doubtless show the art.

10ᵛD *E son lo primo scolar delo magistro che me denanci, e aquesta punta fazo che la ese della sua coverta. Anchora digo che della posta de vera crose, a de posta de crose bastarda po se fare aquesta punta e digo de subito zoe come lo zugadore tra una punta alo magistro o scolar che fosse in le dite guardie, overo poste, lo magistro overo scolar de andare basso cum la persona e pasar fora de strada atraversando la strada, E cum la punta erta al volto overo al petto cum lo mantenir della spada a basso come dipento aqui.*

I'm the previous master's first student, and I make this thrust because it belongs to his cover; also, I say that this thrust could be made from the Position of the True Cross and from the Position of the Bastard Cross.

And I say that as the player quickly throws his point at the master (or student) who was in said guards (or positions), the master (or student) should move lower with his body and step out of the way (across the way), with his point upward at the [player's] face or chest and the cross of his sword held low (as is drawn here).

---

⁸⁹ Interrupted by a sword.

11ʳA *Quando lo scolar che me denanci non po fare lo zugadore cum la punta, e llo acrese lo suo pe stancho dredo lo suo dritto. E lla punta de sua spada gle mette sotto, la gola per butarlo in terra como aqui dipento.*

When the previous student can't finish the player with a thrust, he advances his left foot behind [the player's] right, and he thrusts the point of his sword under the [player's] throat to throw him to the ground (as is drawn here).

11ʳB *Aquesto scolar che denanci de mi sello non po butar lo zugadore in terra cum la punta de sua spada, e cum lo suo pe stancho dredo lo suo dritto, ello passa cum lo pe dritto dredo lo suo stancho, e la spada gle butta al collo. E aquesto zogo io lo fazo ~~icon~~ in suo scambio.*

If this previous student can't throw the player to the ground with the point of his sword and with his left foot outside of [the player's] right, he steps in front of [the player's] left foot with his right and throws his sword around [the player's] neck. And I make this play on his behalf.

11ʳC *A nchora digo io che son lo quarto scolaro che lo nostro magistro po fare aquesto zogo per la coverta che ello ha fatta zoe che ello de passar cum lo pe dritto innanci. E lla punta della sua spada debia meter sotto lo suo dritto brazo. E cum lo brazo stancho seguisca la spada. E quando ello e passato cum lo suo brazo ~~br~~ stancho, lo suo dritto per sotto lo cubito. Ello revolta in la ligadura sottana senza dubito. Quelo che non a fatto per lui lo fazo. Ello porta lo honore e mi lo impazo.*

I, who am the fourth student, also say that our master can make this play from the cover that he has made—that is, that he should step forward with his right foot, and he should thrust the point of his sword under [the player's] right arm, and then follow the sword with his left arm. And when he has passed the [player's] right arm with his left arm behind the [player's] elbow, he'll twist him into the Under Bind without a doubt. That which he hasn't done, I do for him—he'll have the honor, and I the trouble.

11ʳD *Lo mio magistro ma mostrado che quando io son armado. E uno mi vole metere per tale modo la punta in lo volto. Che io debia meter lo taio della mia spada sotto la sua man stancha. E debia passar cum lo pe dritto dredo lo suo stancho, e per tal modo gle poro ferire cum lo pomo overo cum lelzo in lo volto. Come po questo scolar che me dredo far. Sio lo passo fazo tale zogo non me po falire.*

My master has shown me that when I'm armored and someone wants to thrust his point in my face like this, I should thrust the edge of my sword under his left hand and I should step behind his left foot with my right, and in this way, I can strike his face with my pommel or with my elbow, as can this following student. If I can perform such a play, it won't fail me.

Quando lo scolar che m'è denanci non po ferir lo zugadore cum la punta, ello acresse lo suo pe stancho dredo lo suo dritto. E la punta de sua spada gli mette sotto la gola p[er] butarlo interra como aqui depento.

A questo scolar che denanci d'mi s'ello no po butar lo zugadore interra cum la punta de sua spada e cu lo suo pe stancho dredo lo suo dritto. Ello passa cum lo pe dritto dredo lo suo stancho e la spada gle butta al collo. E questo zogo io lo fazo p[er] i' suo scambio.

Anchora digo io che son lo quarto scolaro che lo nostro magistro po fare aquesto zogo p[er] la coverta ch'ello ha fatta, zoè che ello è di passar cum lo pe dritto inanci. E la punta della sua spada debia meter sotto lo suo dritto brazo. E cum lo brazo stancho sognisca la spada. E quando ello è passato cu' lo suo brazo dritto stancho lo suo dritto p[er] sotto lo cubito. Ello va volta in la ligadura sottana senza dubito. Quelo che no è fatto p[er] plu' lo fazo. Ello porta lo honore e mi lo impazo.

L'omo magistro ma mostrado che quando io son armado. E uno mi vole metere p[er] tale mo' la punta i' lo volto. Ch'io debia metere lo taio della mia spada sotto la sua man stancha. E debia passar cum lo pe dritto dredo lo suo stancho e p[er] tal modo gle p[er]zo ferire cu' lo pomo overo cu' l'elzo i' lo volto. Come po' questo scolar che me dredo far. Si o lo passo fazo tale zogo non me po falir.

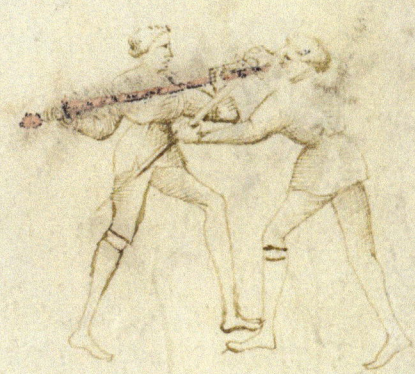

Aquello che disse lo scolaro che denanzi non falla
che io ti posso fare aquesti dui zoghi che dicto
me stano. zoe lo p°mo ferire cu lo pomo in
lo volto. e sbattete in terra. l'altro che segue
che lo segondo, quello ser p° fare acresendo lo
pe ditto e lo mantenir dela spada. e cu illo
lo fiero in la orechia e in la guanza. p tal
modo andera in terra senza dubitanza.

Lo scolaro che denanzi dello magistro e bene in
formado che delo pomo te fiero in lo volto e ti
sbatero riverso in terra. per tale modo che
non ti trovera parazone ne cozulera.

Aquesto zogo non ho falato/ come ha ditto lo
segondo scolaro che denanzi che io possa ve
nire aquesto zogo che te fiero in la orech
ia over la guanza. de mandarte in terra io mi
togno certo. Tu anderai al piano e mi
staro in erto.

dello magistro ne de suoy scolari p echo fazo cum
Questo contrario fazo contra lui cum bona mi
sura. zoe quando ello vene cum la cruzeta e
io cum lo mio stincho p brazo rebato lo suo sti
cho brazo cubito. E p il mo lo cosi posso
defesa non po fare e po essere offeso. Anche
un altro contrario possua fare. Cum la mia
mane stancha p lo sito cubito lo possua voltar.
Che tal zogo in arme e senza arme ben se po
fare.

11ᵛA *Aquello che dise lo scolaro che denanci non falla che io ti posso fare aquesti dui zoghi che dredo me stano, zoe lo primo ferire cum lo pomo in lo volto, e sbaterti in terra. Laltro che segue che lo segondo, quello se po fare acresere lo pe dritto e lo mantenir dela spada, e cum quello lo fiera in la orechia, e in la guanza, per tal modo andera in terra senza dubitanza.*

That which the previous student said doesn't fail me: I could make these next two plays—that is, the first strikes your face with my pommel and knocks you to the ground; the other (which is the second) follows, that if I can advance my right foot and the cross of my sword, then with that, I strike you in the ear and in the cheek, and in that way you'll go to the ground without a doubt.

11ᵛB *Lo scolaro che denanzi dello magistro, e bene informado che delo pomo te fiero in lo volto e ti sbatero riverso in terra, per tale modo che non ti zovara pancirone ne cervilera.*

The master's previous student is well-informed: I strike you in the face with my pommel, and then I knock you to the ground in such a way that neither your breastplate nor your helmet will save you.

11ᵛC *Aquesto zogo non ho falato, come ha ditto lo segondo scolaro che denanci che io posseva venire ad aquesto zogo che te fiero in la orechia e in la guanza, de mandarte in terra io mi tegno certo. Tu andarai al piano e mi staro in erto.*

This play hasn't failed, as the second student before me said, because I could come to this play and strike you in the ear and in the cheek, and I hold myself certain of sending you to the ground. You'll go to the ground and I'll remain upright.

11ᵛD *Dello magistro ne de suoij scolari pocho fazo cura. Questo contrario fazo contra lu cum bona misura, zoe quando ello vene cum la coverta, e io cum lo mio stancho brazo rebato lo suo stancho brazo cubito. E per tal mo cosi preso defesa non po fare e po essere offeso. Anchora altro contrario poseva fare. Cum la mia mane stancha per lo suo cubito lo posseva voltar. Che tal zogo in arme e senza arme ben se po fare.*

I care little for the master or his students. I do this counter against him for good measure: that is, when he comes with the cover, I deflect the elbow of his left arm with my left arm, and because of this hold, he can make no defense and he can be harmed.

I could do another counter, too: I could turn his elbow with my left hand. Such a play can be done well both in armor and without.

12ʳA *Aqui comenzano le guardie de spada a do mane e sono ·XII· guardie, e la prima si e tuta porta di fero che sta in grande forteza e si e bona da aspetar ogni arma manuale longa, e curta, e pur chello abia bona spada non cura de tropa longheza. Ella passa cum coverte e va ale strette. Ella scambia le punte e le soi ella mette. Anchora rebatte le punte a terra, e sempre va cum passi e de ogni colpo ella fa coverta. E chi in quella gle da briga grande defese fa senza fadigₐ.*

Here begin the guards of the sword in two hands, and there are 12 guards.

The first is the Full Iron Gate, which stands in great strength and is good for awaiting all hand-held weapons, both long and short, and if she has a good sword then she doesn't care if it's very long. She steps with a cover and goes to the constrained play; she exchanges thrusts and she delivers her own; she beats thrusts to the ground, too, and always goes with steps and makes covers against all blows. And she'll make great defense against anyone who starts a fight with her.

12ʳB *Questa si e posta de donna che po far tuti setti colpi dela spada. E de tuti colpi ella se po covrir. E rompe le altre guardie per grandi colpi che po fare. E per scambiar una punta ella e sempre presta. Lo pe denanci acrese fora de strada, e aquello de dredo passa ala traversa. E llo compagno fa romagner discoverto. E aquello po ferire subito per certo.*

This is the Position of the Noblewoman, which can make all seven blows of the sword and can also cover all blows. She breaks the other guards by making great blows, and she is always ready for the exchange of the thrust: the front foot advances out of the way and the rear one steps to the side, so her companion is left uncovered and she can certainly strike him quickly.

12ʳC *Questa si e posta de fenestra dextra che de malicie e inganni sempre e presta, E de covrir e de ferir e lla magistra. E cum tute guardie ella fa questione, e cum le soprane, e cum le terene. E duna guardia a laltra ella va spesso per inganare lo compagno. E a meter grande punte e saverle rompere e scambiar. Quelli zoghi ella po bene fare.*

This is the Position of the Window on the right, which is always ready with malice and trickery, and she's the master of covering and of striking, and she makes her dispute with all the guards (with the high and with the low). She often goes from one guard to another in order to fool her companion, and she throws great thrusts and knows how to break and to exchange them—she can make those plays very well.

12ʳD *Questa e meza porta di ferro perche sta in mezo, e de una forte guardia ma ella non vole longa spada, ella zetta forte punte, e rebatte per forza le spade in erto, e torna cum lo fendent per la testa o per gli brazi, e pure torna in sua guardia. Per zo vene chiamada porta, per che ella e forte. E de forte guardia che male se po rompere senza pericolo a venir ale strette.*

This is the Middle Iron Gate (because it stands in the middle) and she's a strong guard, but she doesn't want a long sword. She throws strong thrusts and beats swords upward with strength, returns with a downward blow for his head or for his arms, and then simply returns to her guard. She's well-named "gate" because she's strong: she's a strong guard that can't easily be broken without danger and without coming to the tight [play].

Aqui comenzano le guardie de spada ad o mane e sone .xij. guardie e la prima sie tuta porta di fero che sta in grande forteza e sie bona da aspetar ogni arma manuale longa e curta, e pur ch'ella sia d'una bona spada non cura de tropa longheza. Ella passa cum coverte e va ale strette. Ella scambia le punte e le soi ella mette. Anchora za rebatte le punte a terra e sempre va cu passi e de ogni colpo ella fa coverta. E chi con ella a che far e da briga grande defese fa senza fadiga.

Questa sie posta de dona che po far tuti setti colpi dela spada. E de tuti colpi ella se po covrir e rompe le altre guardie per grandi colpi ch'ella po fare. E per scambiar una punta ella se sempre presta. Lo po dar una acresse forza de strada e a quello de drieto passa dela traversa. E lo compagno fara zo magnir discoverto. E a quello po ferire subito porto.

Questa sie posta de fenestra dextra che de malicie e inganni sempre e presta. E de covrir e de ferir ella magistra. E cum tute guardie ella fa questione e cum le soprane e cun le terzane. E d'una guardia a l'altra ella va spesso per ingannare lo compagno. E a meter grande punte e savere rompe e scambiar, quelli zoghi ella po bene fare.

Questa e meza porta di fero, per che sta in mezo e de una forte guardia, ma ella no vole longa spada. Ella zetta forte punte e rebatte per forza le spade in erto e torna cu lo fendenti per la testa o per gli brazi e pur torna in sua guardia. E zo vene chiamada porta per che ella e forte e forte guardia che male se po rompe senza pericolo a venir ale strette.

Questa sie posta longa che piena de falsita. E-
lla va tastando le guardie se lo compagno po inganar. Sella po ferire de punta ella lo sa ben
fare. gli colpi ella schiva e poi feriss sella lo po
fare. piu che le altre guardie le falsitade sa
vsare.

Questa sie posta frontale e alchun magistri la
chiamano posta di corona. Che pi[n] crosar ella
e bona e p le punte ella e anchora bona che se
la punta gli vene trata erta ella la incrosa passa
do fora de strada. E sella punta si ven trata bassa
anchora passa fora de strada rebatendo la punta al[er]
za anchora po fare altramente. Che i[n] lo trar dela punta torni cum lo pe i[n] dredo e vegna discendente p[er]
la testa e p gli brazzi e vada in dreto dezegna-
ro e subito bute vna punta o da cu[m] acresc[er]
de pe c[on]tra guardia.

Questa sie posta de donna che p[ro] tuti gli sette colpi fare della spada fare. E de tuti colpi ella
se po covrire. E rompe le altre guardie p
grandi colpi che po fare. E p scambiar v-
na punta ella e sempre presta. Lo pe che
denanci acreso fora de strada, e aquello de dre
do passa ala trauersa. Elo compagno fa
romagnare discouerto. E aquello po ferir
subito p certo.

Questo sie dente de zenghiar che dello dente de
zenghiaro fa simile modo de ferir. Ello tra
grande punte p sotto le mane in fin al volto e
non se moue da passo e torna cum lo feridente
zo p li brazzi. E alchuna volta tra la punta al
volto e da cui la punta erta e si quello butar de
punta ello acrese lo pe che denaci subito e tre-
na cui lo feridente p la testa e p li brazzi e torna
in sua guardia e subito zetta vnaltra punta
cu[m] accrestere de pe e ben se defende dalo zogo streto.

12ᵛA *Questa si e posta longa che piena de falsita. Ella va tastando le guardie se lo compagno po inganar. Sella po ferire de punta ella lo sa ben fare, gli colpi ella schiva e poi fiere sella lo po fare. Piu che le altre guardie le falsitade sa usare.*

This is the deceitful Extended Position; she probes the guards to see if she can deceive her companion. If she can strike with a thrust, she knows how to do it well; she dodges blows and she can wound when she's able. More than any other guard, her tactic is deception.

12ᵛB *Questa si e posta frontale e alchuni magistri la chiamano posta di corona. Che per incrosar ella e bona, e per le punte, ella e anchora bona che se la punta gle vene trata erta ella la incrosa passando fora de strada. E se lla punta si ven trata bassa anchora passa fora de strada rebatendo la punta a terra. Anchora po fare altramente, Che in lo trar dela punta torni cum lo pe indredo, e vegna di fendenti per la testa e per gli brazi e vada in dente do zenghiaro e subito[90] bute una punta, o doe cum acreser de pe, e torni[91] di fendent cum quella propria guardia.*

This is the Headband Position,[92] which some masters call the Position of the Crown. She's good for crossing and for thrusting, and is also good if a point comes attacking upwards: she crosses, stepping out of the way. And she also steps out of the way if a point comes attacking low, beating the point to the ground.

She can do it differently, too, such that in the attack of the point, she steps backward with her foot and comes with a downward blow for his head and for his arms and goes to the Wild Boar's Tusk, and then quickly throws a thrust or two with an advance of her foot and returns with a downward blow from her own guard.

12ᵛC *Questa si e posta de donna che po tuti gli setti colpi fare della spada fare. E de tuti colpi ella se po covrire. E rompe le altre guardie per grandi colpi che po fare. E per scambiar una punta ella e sempre presta. Lo pe che denanci acrese fora de strada, e aquello de dredo passa ala traversa. E lo compagno fa romagnire discoverto. E aquello po ferir subito per certo.*

This is the Position of the Noblewoman, which can make all seven blows of the sword and can also cover all blows. She breaks the other guards by making great blows, and she's always ready for the exchange of the thrust: the front foot advances out of the way and the rear one steps to the side, so her companion is left uncovered and she can certainly strike him quickly.

12ᵛD *Questo si e dente de zenghiar, che dello dente de zenghiar fa simile modo de ferir. Ello tra grande punte, per sotto le mane in fin al volto, e non se move de passo, e torna cum lo fendente, zo per li brazi. E alchuna volta tra la punta al volto, e va cum la punta erta, e in quello butar de punta, ello acrese lo pe che denanzi subito, e torna cum lo fendente per la testa, e per li brazi, e torna in sua guardia, e subito zetta unaltra punta cum acresere de pe, e ben se defende dalo zogo streto.*

This is the Wild Boar's Tusk, since the tusk of a wild boar has a similar method of striking. She throws great underhand thrusts ending in his face and doesn't move a step, and then returns with a downward blow down at his arms. And sometimes she throws her point at his face and goes with the point high, and she quickly advances her front foot during that throw of the point and returns to her guard, and then throws another quick thrust with an advance of her foot. And she protects from the tight play well.

---

[90] Word bisected by sword.
[91] Word bisected by sword.
[92] *Frontale* as a noun is a type of headband, whereas as an adjective it simply means "frontal". *Posta frontale* could be read literally as "frontal position", but it could also be a shortened form of *posta di frontale*, which is how I translate it here because that seems to fit the overall naming pattern better. A similar elision of the *di* in *posta di donna* can be seen on folio 4ᵛB.

13ʳA *Questa si e posta breve che vole longa spada, ed e una maliciosa guarda che non ha stabilita. Anche sempre se move, e guarda se po intrar cum punta e con passo contra lo compagno, e piu e apropiada tal guardia in arme che senza arme.*

This is the Short Position, which wants a long sword, and she's a malicious guard which has no stability. Also, she always moves and watches to see if she can enter against her companion with a step and a thrust. This guard is more appropriate in armor than without armor.

13ʳB *Questa si e posta de donna la sinistra che de coverte e de ferire ella e sempre presta. ella fa grandi li[93] colpi e rompe le punte, sbattele a terra. E intra in lo zogo stretto, per lo saver atravesare. Aquisti zoghi tal guardia sa bene fare.*

This is the Position of the Noblewoman on the left, and she's always ready to cover and to strike. She makes great blows and breaks thrusts, beating them to the ground, and knowing how to thwart, she enters into the tight play. This guard knows how to make these plays well.

13ʳC *Questa si e posta de coda longa che distesa in terra de dredo ella po mettere punta, e denanci po covrir e ferir. E sello passa innanci e tra delo fendente. In lo zogo stretto intra senza falimento. Che tale guardia e bona per aspetare. Che de quella in le altre tosto po intrare.*

This is the Position of the Long Tail, which is extended toward the ground. She can thrust the point from behind and she can cover and strike in front, and if she steps forward and throws a downward blow, she enters into the tight play without failure. Such a guard is good for waiting, because from it she can readily enter into the others.

13ʳD *Questa e posta de bicorno che sta cosi serada che sempre sta cum la punta per mezo la strada E aquello che po fare posta longa po far questa E similmente dico de posta di fenestra, e posta frontale.*

This is the Position of the Two-Horned Anvil,[94] which is so strongly enclosed that she always remains with her point toward the middle of the way. And she can do that which the Extended Position can do, and this can similarly be said about the Position of the Window and the Headband Position.

---

[93] Corrected from "e" to "i".

[94] A *bicorno* is a type of anvil with a horn on either end. I think this is the imagery being invoked with this name, but it could alternatively be read in a generic sense as "the position of the two-horned thing".

Questa sie posta breue che vole longa spada, e de una maliciosa guarda che non ha stabilita. Anchi sempre se move, e guarda se po intrar cum punta e con passo contra lo compagno. E piu e aproprida tal guardia in arme che senza arme.

Questa sie posta de dona la sinistra che de coverte e de ferire ella e sempre presta. Ella fa grandi colpi, e rompe le punte, e bate le a terra. E intra i lo zogo stretto, p lo sauer atrauersare. A questi zoghi in tal guardia sa bene fare.

Questa sie posta de coda longa che distesa intesa de drieto ella po mettere punta, e denanci po couvrir e ferire. E sello passa inanci, e tra delo fendente. In lo zogo stretto intra senza fallimento. Che tale guardia e bona p aspettare. Che de quella in le altre tosto po intrare.

Questa e posta de bicorno che sta cosi serada che sempre sta cum la punta p mezo la strada. E aquello che po fare posta longa, po far questa. E similmente dico de posta di fenestra e di posta frontale.

Aqui comenza zogo largo de spada a doe mane cum pcho incrosare
lo honore sera de chi meio savera fare.

Questo magistro che aqui incrosado cum questo
zugadore dise quando io sono incrosado in punta de spada subito io do volta a la mia spada e si lo fiero de l'altra pte cum lo fendente zoe p la testa e
p gli brazzi ouero cho gle metto una punta
in lo volto. Como uuy vederite qui de drieto
da my dipento.

Io te ho posto una punta in lo volto come lo mio
magistro che denanci dise. Anchora pma auere
fato zo ch'ello dise zoe auere trato de mia spada
subito quando io era apresso lo incrosare della pte
dritta dell'altra pte zoe della stancha io debeua uoltare subito la mia spada in lo fendente p la testa
e p gli brazzi come ha ditto lo magistro mio ch
denanci.

Anchora me incroso qui p zogo largo a meza spada. E subito quando io sono incrosado io lasso
discorere la mia spada sopra le sue mane, e se uo
glo passare cum lo pe dritto fora de strada io g
li posso mettere una punta in lo petto come uuy
vedete qui drieto dipento.

Io zogho de lo mio magistro io lo compido che
io fatto la sua couerta. E subito o fatto el
suo ditto. Ch'io feridi pma gli brazzi e poy
gle ho posta la punta e lo petto.

13ᵛ *Aqui comenza zogo largo de spada a doe mane cum pocho incrosare, lo honore sera de chi meio savera fare.*

The loose play[95] of the sword in two hands begins here with a little crossing; the honor will belong to whomever will know how to do it.

13ᵛA *Questo magistro che aqui incrosado cum questo çugadore dise quando io sono incrosado in punta de spada subito io do volta ala mia spada, e si lo fiero de laltra p<u>ar</u>te cum lo fendente zoe p<u>er</u> la testa, e p<u>er</u> gli brazi, overo o chio gli metto una punta in lo volto. Como vuij vederite qui de dredo da mi dipento.*

This master who's crossed here with this player says, "When I'm crossed at the tip of the sword, I give a quick turn to my sword and thus I strike with a downward blow from the other side (that is, through his head and through his arms), or I thrust my point into his face as you see hereafter in my depiction."

13ᵛB *Io te ho posto una punta in lo volto, como lo mio magistro che denanci dise. Anchora poria avere fato zo chello dise, zoe avere trato de mia spada, subito quando io era apresso lo incrosare della p<u>ar</u>te dritta dellaltra p<u>ar</u>te zoe della stancha io debeva voltare subito la mia spada in lo fendente p<u>er</u> ~~in~~ la testa e p<u>er</u> gli brazi como ha ditto lo magistro mio ch<u>e</u> denanci.*

I've set my point in his face, as my master (who came earlier) said. I could've made the [other] play that he said, too—that is, to have quickly attacked with my sword when I was near the crossing of the right side: I should've immediately turned my sword into a downward blow from the other side (that is, from the left) through his head and through his arms, as my master (who came earlier) has said.

13ᵛC *Anchora me incroso qui p<u>er</u> zogo largo, a meza spada. E subito quando io sono incrosado io lasso discorere la mia spada sopra le suie mane, e se voglio passare cum lo pe dritto fora de strada io gli posso mettere una punta in lo petto come vui vedete qui dredo dipento.*

I'm crossed here at mid-sword, again for the loose play. And when I'm crossed, I quickly allow my sword to run off over his hands, and if I want to step out of the way with my right foot, I can thrust my point into his chest (as is drawn hereafter).

13ᵛD *Lo zogho delo mio magistro io lo compido, che io fatto la sua coverta. E subito, o fatto el suo ditto. Che io feridi p<u>ri</u>ma gli brazi, e poij gle ho posta la punta i<u>n</u> lo petto.*

I've completed my master's play in that I've made his cover and I've quickly executed his saying: I've struck his arms first, and then I've set my point in his chest.

---

[95] This is typically rendered "wide play", but GUY WINDSOR has correctly pointed out that *stretto* and *largo* have stronger connotations of tight/constrained vs. loose/unrestricted. I have thus translated *largo* as "loose" and *stretto* as "tight". Because these English adjectives often can't be used as stand-alone nouns in quite the same way the Italian words can, I've inserted [play] in a few places to make better sentences.

14ʳA *Ello mio magistro che me denanci ma insegnato che*[96] *quando a meza spada io son cum uno incrosado che subito me debia acresere in denanci e piglare la sua spada a questo partito, per ferirlo taglo, o punta. Anchora gle posso guastare la gamba per lo modo che voi possete vedere aqui dipento a ferirlo cum pe sopra la schena della gamba, overo sotto lo zinochio.*

My master (who came earlier) has taught me that when I'm crossed at mid-sword, I should quickly advance forward and grab his sword (as in this match) in order to strike him with edge or point. I can also wreck his leg in the way that you'll be able to see drawn hereafter by striking with my foot over the back of his leg (or under his knee).

14ʳB *Lo scolar che me denanzi dise del suo magistro, e mio che llo gle ha insignado questo zogo, e per vezuda io lo fazo. A farlo senza dubio ello me pocho impazo.*

About his master and mine, the previous student says that he has taught this play, and I do it to crumple [my opponent]. Without a doubt, doing it is little trouble to me.

14ʳC *Questo zogo e chiamado lo colpo del vilano, e si fa per tal modo. Zoe che si de aspetare lo vilano chello traga cum sua spada. E quello che aspeta lo colpo di stare in picolo passo cum lo pe stancho denanci. E in quello che lo vilano te tra per ferirte. Acrese lo pe stancho fora de strada inverso la parte dritta. E cum lo dritto pe passa ala traversa fora de strada, piglando lo suo colpo a meza spada e lassa discorere la sua spada a terra, e subito respondigli cum lo fendente per la testa overo per gli brazi overo cum la punta in lo petto come qui dipento. ancora questo e bono.*

This play is called the Peasant's Strike, and it's made in this way (that is, that one should await the peasant like this until he strikes with his sword), and he who awaits the blow should stand in a small stance with his left foot forward. And in that moment when the peasant throws his strike at you, advance your left foot out of the way against his right side. And step out of the way to the side with your right foot, catching his blow at mid-sword and allowing his sword to run off toward the ground, and then quickly respond with a downward blow (through his head or through his arms), or with your point in his chest (as drawn here)—this is good, too.

14ʳD *Questo e lo colpo del vilano che qui denanci de mi. Che bene gle ho posta la punta in lo petto. E cosi gle posseva fare uno colpo per la testa, e per gli brazi cum lo fendente como e ditto denanci. Ancora seio[!] volesse lo zugadore contra de mi fare che ello volesse ferirme cum lo riverso sotto gli mie brazi. Io subito acrescho lo pe stancho e meto la mia spada sopra la sua, e non mi po far niente.*

This is the peasant's strike (which appeared here previously), so I've put my point into his chest well. And I could [also] have made a cut through his head and through his arms with a downward blow (as was said previously).

Also, if the player wanted to come against me such that he would strike me with a backhand blow under my arms, I would quickly advance my left foot and thrust my sword over his, and then he can't do anything to me.

---

[96] *ma insegnato che* partially effaced.

Ello mio magistro che me denanzi me insignado che
quando a meza spada io son cum uno incrosado che
subito me debia acrescere in denanzi e pigliare la sua
spada a questo modo. Per ferirlo taglio o punta
in volta gle posso guastare la gamba per lo modo
che voi possete vedere aqui dipento. E ferirlo cum pe
sopra la schena della gamba overo sotto lo zinochio.

Lo scolar che me denanzi disse del suo magistro e mio
che lo gle ha insignado questo zogo e questa
io lo fazo. A farlo senza dubio ello me specto per
pazo.

Questo zogo e chiamado lo colpo del vilano e
si fa per tal modo. Zoe che si de aspettare lo vilano
che lo traga cum sua spada. E quello che aspetta
lo colpo di stare in pocolo passo cum lo pe stancho
denanzi. E in quello che lo vilano te tra per ferirte
il crese lo pe stancho fora de strada in verso la parte
dritta. E cum lo dritto pe passi ala traversa fora
de strada e pigliando lo suo colpo a meza spada e la
sa discovrire la sua spada aterra e subito respondigli
cum lo fendente per la testa overo per gli brazzi overo cum la
punta in lo petto come e dipento. Ancora questo e bono.

Questo e lo colpo del vilano che qui denanzi dessi
che bene gle lo posta la punta in lo petto. E
cosi gle possena fare uno colpo per la testa e per gli
brazzi cum lo fendente como e ditto denanzi. Ancora
za scio volesse lo zugadore contra de mi fiere che
ello volesse ferirme cum lo riverso sotto gli mie
brazzi. Io subito acresco lo pe stancho e meto
la mia spada sopra la sua e no mi po fer niente.

Quando uno te tra p[er] la gamba, discresse lo pe che denanci. O tu lo torna indredo e tu dello ferendi p[er] la sua testa. Con[m]e aqui dipento. Ben che cu[m] spada a doe mane non se de' traire dello genochio in zu. Però è tropo grande p[er]icolo a quello che tra. Che lo romane tuto discou[er]to quello che tra p[er] la gamba. Salvo che se uno fosse cu[m]zu[n]to intera[m]e[n]te, ben se poria trar p[er] gamba. Ma altramente non stando spada contra spada.

A questo partito che io te fiero cu[m] lo p[unt]e p[er] gli ochi, el fazo p[er] farte do[l]a, e p[er] farte suariarte la covertu[n]a, che fazando questo zogo vole essere fato subito, e p[er] non auer dello contra[r]io dubito. Lo contrario de questo zogo vole essere p[re]sto fato zoeche lo zugidore de pigliare p[er] la gamba dritta lo scolaro cu[m] sua mane stancha e i[n] terra lo po' butare.

Questo zogo che se chiama scambiare de punta, e se fa p[er] tal modo zoè quando te tra una punta subito acresse lo tuo pe denanci fora de strada e cum l'altro pe passa a la traversa anchora fora de strada, atraversando la sua spada cum gli to[y] braz[z]i bassi. E cum la punta de la spada e[n]tra in lo volto o in lo petto co[n]me è aqui dipento.

De questo scambiar de punta che m[e] de[n]anci e[ss]e a questo zogo. Che subito lo che lo scolaro e me de[n]anci non metesse la punta i[n] lo volto de lo zugadore chello la falasse chello no' la metesse i[n] lo b[ra]zo o n[e]l lo petto o che fosse lo zugadore armato. Subito debia lo scolar cu[m] lo pe stancho i[n] nanci passar. E p[er] questo modo lo debia pigliare e la sua spada metero a bon ferir. P[er]ò che lo zugadore e' p[re]sa su[a] spada no' po' fuzir.

14ᵛA *Quando uno te tra per la gamba, discrese lo pe che denanci. O tu lo torna indredo e tra dello fendent per la sua testa. Come aqui dipento. Ben che cum spada a doe mane non se de trare dello genochio in zu, perche e tropo grande pericolo aquello che tra. Che lo romane tuto discoverto quello che tra per la gamba. Salvo che se uno fosse cazuto in terra, ben se poria trar per gamba. Ma altramente non stando spada contra spada.*

When someone throws [a strike] at your leg, withdraw your forward foot (or return it behind), and throw a downward blow at his head (as drawn here). Note that the sword in two hands shouldn't attack below the knee because it's too dangerous for the attacker: he that attacks the leg remains wholly uncovered, unless he'd fallen to the ground—then he could strike the leg well, but otherwise [he could] not when fighting sword against sword.

14ᵛB *Aquesto partito che io te fiero cum lo pe in gli coglura el fazo per farte doia, e per farte svariarte la coverta che fazando questo zogo vole essere fato subito, per non aver dello contrario dubito. Lo contrario de questo zogo vole essere presto fatto zoe che lo zugadore de piglare per la gamba dritta lo scolaro, cum sua mane stancha e in terra lo po butare.*

This match is where I strike you in the testicles with my foot, and I do it to give you pain and to make your cover waver, so in making this play, I want to do it quickly so that your counter is doubtful.

The counter to this play wants to be done quickly, which is that the player should grab the student by the right leg with his left hand, and then he can throw him to the ground.

14ᵛC *Questo zogo che se chiama scambiare de punta, e se fa per tal modo zoe quando te tra una punta subito acrese lo tuo pe denanci fora de strada a cum laltro pe passa ala traversa anchora fora de strada, atraversando la sua spada cum gli toij brazi bassi, E cum la punta dela spada erta in lo volto o in lo petto come e aqui dipento.*

This play, which is called the Exchange of the Thrust, is made like this: that is, that when he throws his point at you, quickly advance your forward foot out of the way and step to the side with your other foot (also out of the way), crossing his sword with your arms low and with the point of your sword up in his face or in his chest (as is drawn here).

14ᵛD *De questo scambiar de punta che me denanci essi aquesto zogo. Che subito lo che lo scolar che me denanci non metesse la punta in lo volto delo zugadore chello la falasse chello non la metesse in lo volto ne in lo petto, o perche fosse lo zugadore armato. Subito debia lo scolar cum lo pe stancho innanci passar. E per questo modo lo debia piglare, e la sua spada metero[!] a bon ferire, po che lo zugadore e presa sua⁹⁷ spada non po fuzir.*

From this earlier exchange of the thrust comes this play. Given that the previous student didn't immediately thrust his point into the player's face (or that he failed such that he couldn't thrust into his face or into his chest, or that the player was armored), then the student should quickly step forward with his left foot and he should grab [the player] like this. And [the student] will throw a good strike with his sword, because the player is held by his sword and can't flee.

---

⁹⁷ Word bisected by sword.

The Flower of Battle: MS M 383

16ʳ *E qui comenza zogo de spada a doi mane, zogo stretto. El modo de rompe tute punte e tagli. In lo quale serano dogni rasone, coverte, feride, e ligadure, e dislogadure, e prese, e tore de spada, e sbatere in terra, per diversi modi. E serano gli remedij, e gli contrarij de zaschuna rasone che bisogna a offendere e difendere.*

Here begins the play of the sword in two hands, the tight play, the method of breaking all thrusts and cuts, in which will be every type of covering and striking, binding and dislocations, holds and takings of the sword, and knocking to the ground in diverse ways. And there will be every type of remedy and counter that should offend or defend.

16ʳA *Nui stasemo qui incrosadi e di questo incrosare che noij fazemo. Tuti gli zoghi che noij segueno fargle possemo. Acosi uno de noij, quale laltro. E tuti gli zoghi seguirano uno laltro como denanci e dito.*

We stand here crossed and from this crossing that we do, we can make all the plays that follow us (either one of us as [easily] as the other). And all these plays will follow, one after the other, as was previously said.

16ʳB *Per lo incrosar che a fatto lo magistro cum lo pe dritto denanci io comprischo lo primo zogo, zoe che io passo cum lo pe stancho, e cum la mia mane stancha passo di sopra lo mio dritto brazo, e piglo lo suo mantenir dela spada sua, in mezo le soe[!] mane zoe in mezo delo mantenir. E cum taio e punta, Io lo posso ferir, e questa presa se po fare cosi a spada de una mane, come a doe mane, e cosi a incrosar de sotto mane come di sopra se po far tal presa.*

From the crossing that the master has made with his right foot forward, I complete the first play—that is, I step with my left foot, and I pass my left hand over my right arm and grasp the hilt of his sword in between his hands (in the middle of the hilt), and then I could strike him with edge and point.

This hold can be made as easily with the sword in one hand as with two hands, and this hold can be made as easily crossing under the hands as over.

16ʳC *Questo e uno altro zogo che vene delo incrosar dello mio magistro. E como ello e incrosado ello po fare questo zogo e lli altri ~~po~~ che segueno de dredo, zoe chello po fare overo piglare lo çugadore a questo modo, e ferirlo in lo volto cum lo pomo de sua spada. Anchora po ferirlo de fendente in la testa. Innanci chello fare coverta presta.*

This is another play that comes from my master's crossing, and since he is crossed, he can make this play and the others that follow after—that is, he can do or grasp the player like this to strike him in the face with the pommel of his sword. He can also strike him in the head with a downward blow before [the player] could make a cover ready.

16ʳD *Questo e uno altro ferir de pomo e se po far subito, se lo volto e discoverto fallo senza dubito, Che lo se po fare armado e disarmado. Quatro denti fora butta de boccha a uno cum tale zogo, che la provado. E lla spada se volesse al collo te poria butar come fa dredo da mi quello scolaro.*

This is another strike of the pommel and it can be done quickly; if his face is uncovered then do it without doubt, because it may be done armored or unarmored. You'll knock four teeth from his mouth with this play (as has been proven), and if you wanted, you could throw your sword around his neck as the next student does.

Equi comenza zogo de spada a dui mane zogo stretto. El modo de rompe tute punte e tagli. In lo quale serano d'ogni rasone coverte ferite e ligadure e dislogadure e prese e tore de spada e sbattere interim p diversi modi. E seza no gli remedy e gli contrary de zaschuna rasone che bisogna a offendere e difendere.

Nui stassemo qui mezo spada e di questo in cosi fare che noy fazemo tutti gli zoghi che noy seguemo fargle possemo. A cosi uno de noy quale l'altro. E tutti gle zoghi seguiramo uno l'altro come denanci e ditto.

Per lo incrosar che a fatto lo magistro cu lo predutto de nanzi io compischo lo primo zogo zoe che io passo cu lo pe stancho e cu la mia mane stancha passo di sopra lo mio ditto brazo e piglio lo suo mantenir de la spada siu in mezo le soe mane zoe in mezo de lo mantenir. E cu tanto e putta io lo posso ferir e questi presi se po fare cosi a spada da una mane come a due mane e cosi a incrosar de sotto mane come di sopra se po far tal presa.

Questo e uno altro zogo che vene de lo incrosar de lo mio magistro. E como ello e incrosado ello po fare questo zogo elli altri po che seguono dedredo. Zoe che llo po fare uno pigliare lo suo zoguidor in questo modo e scurlo in lo volto cu lo pomo de sua spada. Anchora po scurlo de fendente in la testa. In nanza che llo fare coverta p(re)sta.

Questo e uno altro ferir de pomo e se po far subito solo volto e stordido fallo senza dubito. Che lo se po fare armado e disarmado. Quatro denti fora butta de bocha e uno cu tale zogo che la provado. E la spada si volesse al colo te poria butar come fa dedro arm(e) quello scolaro.

pa lo zogo che me denanzi e como lo scolar ha
ditto io t'o posta la spada al collo. E la gola
te posso ben taglar. p(er)ço chio sonto che tu
non ai punto de collaro.

Quando io sonto incrosado io passo cu(m) coverta
E fiero i(n) gli tovi brazi a questo p(ar)tito. E questa
punta e pegno in lo to volto. E silo p(er) stancho
io acresco in tra(m)be le braze ti ligaro. Overo di
questo zogo ch(e) m(e) dredo te piglio zo
che te ligaro la spada e p(er) lezo la tegniro

La presa che dise lo scolar che m(e) denanzi quella ti faro ferir te posso senza i(m)pazo e tegno
lezo de tua spada de punte e tagli ti fiero de
rida. E questo zogo remi(n) ogni toze de
spada allo zogo stretto subito fralo quello giu-
sta.

Questo sie p(er) altro contrario che co(n)tra lo
toze la spada E fo si p(re)sto e subito p(er) tale
modo. Quando uno si vene troga a mezzo
spada o acresse stretto cu(m) lo pe sta-
che de nanzi e la s(u)a mano stancha pigla
la s(u)a mane in la zuntura della sua mane
p(er) sotto lo so ma(n)ter(er). E volta la sua spada i(n) vesta p(ar)te reversa p(er) tale modo alzaq
d(e) p(ar)te. E silo fieri in lo p(et)to o i(n) lo t(er)ra
o vero li ch(e) mego gle p(ar)se.

16ᵛA *Per lo zogo che me denanci, e como lo scolar ha ditto io to posta la spada al collo. E lla gola te posso ben taglar. Per o chio sento che tu non ai punto di colaro.*

From the previous play (and as the student has said), I've placed my sword around your neck, and I can cut your throat well because I feel that you don't have an armored collar.

16ᵛB *Quando io sono incrosado io passo cum coverta E fiero in gli toij brazi a questo partito. E questa punta pogno in lo to[!] volto. E si lo pe stancho io acresco trambe le brazi ti ligaro. Overo che in questo zogo che me dredo te pigliro zoe che te ligaro la spada, e per lelzo la tegnero.*

When I'm crossed, I step with a cover and hurt your arms[98] as shown, then I put this point in your face. If I advance my left foot, I can bind both your arms, or I'll grab you in this next play—that is, I'll bind your sword and hold it by the hilt.

16ᵛC *La presa che dise lo scolar che me denanci quella ti fazo, ferir te posso senza impazo e tegno lelzo de tua spada, de punte e tagli ti faro derada. E questo zogo rompe ogni tore de spada e llo zogo stretto subito farlo quello guasta.*

I hold you the way that the previous student said. I can strike you without trouble and I hold the hilt of your sword: I strike bargain with my point and edge. And this play breaks all takings of the sword, and it spoils the tight play if done quickly.

16ᵛD *Questo si e uno altro contrario che contra lo tore de spada e fe[!] fa presto e subito per tale modo. Quando uno si vene tego a incrosare, o passa, o acrese stretto cum lo pe stancho denanci, e la tua mane stancha pigla la sua mane in la zuntura della sua mane per sotto lo so mantenir. E volta la sua spada in verso parte riversa per tale modo che aqui dipento. E si lo fieri in lo petto, o in lo ventre o vero li che meio[!] gle piase.*

This is another counter against the taking of the sword, and it may be done quickly and readily like this: when someone comes to cross with you and steps or advances to the tight play with his left foot forward, grasp his hand at the wrist with your left hand (under his hilt) and turn his sword toward your left side in the way that's drawn here. And thus you strike him in the chest or in the belly (or wherever he likes least).

---

[98] The text says "I hurt your arms", but the move is a bind.

15ʳA *Questo e lo tore de spada lo sovrano cum lo man-tenir de mia spada io penzo innanci, e cum la man mancha io stringo li⁹⁹ suoij brazi per modo chello conven perdere la spada. E poi de grandi feride gle faro derada lo scolar che me dredo, aquesto zogo mostra como la spada dello zugadore e in terra posta*

This is taking the sword from above: I push forward with the hilt of my sword and with my left hand, I grasp his arms with my left hand in such a way that he must lose his sword. Then I'll strike a bargain with great strikes.

The next student demonstrates this play as the player's sword is positioned on the ground.

15ʳB *Per la presa dello scolar che me denanci de mi a fata la spada in terra te caduda. Tu lo poij sentir Asai feride te posso fare senza mentire.*

The previous student's hold has caused your sword to fall to the ground. You can feel that I could thus make you truly wounded.

15ʳC *Questo zogo se fa per tal modo zoe che uno vada cum lo colpo mezano contra lo mezano de parte riversa e subito vada cum coverta ale strette. E buta la sua spada alo collo dello compagno, piglando la sua mane dritta cum la sua stancha de si in stesso come aqui dipento. Butarlo po in terra senza falimento metendo lo suo pe dritto dredo lo suo dritto.*

This play is done like this: that is, that someone goes with a middle blow against a left-side middle blow, and then quickly goes to the tight [play] (with a cover). He throws his sword around his companion's neck, at the same time grasping his [companion's] right hand with his left (as you see drawn here). Thrusting his right foot behind [his companion's] right, he can then throw him to the ground without fail.¹⁰⁰

15ʳD *Questa e una altra presa de butar uno in terra cum tuta la spada. Zoe che aquesto scolar incrosa cum lo zugadore della parte dritta e passa ale strete, e cum la mane sua stancha penze lo cubitto dritto del zugadore. E subito gle butta la spada al collo piglando la sua propia spada al mezo. Ello suo dritto pe dredo lo suo dritto dello zugadore. Acosi lo butta¹⁰¹ in terra cum pocho honore.*

This is another hold to throw someone to the ground, sword and all: that is, that this student crosses with the player on the right side and steps into the tight [play]; he pinches the player's right elbow with his left hand, and then he quickly throws his sword around [the player's] neck, grasping his own sword in the middle (his right foot behind the player's right). Thus, he throws [the player] to the ground with little honor.

---

⁹⁹ Corrected from *o* to *i*.
¹⁰⁰ There is a discrepancy between 15ʳCD and the equivalent play in the Getty (30ʳB). The Getty illustration resembles 15ʳD, but the text is comparable to 15ʳC (and the Getty doesn't have the other illustration or description). Either manuscript *could* be in error, but the brief descriptions of both plays in the Pisani Dossi support the idea that the error is in the Morgan and the text descriptions should be switched.
¹⁰¹ Word bisected by sword.

Questo e lo toze de spada lo sovrano cum lo matenir de tutta spada io zonzo franch̃ e cum la man
mancha io tenero la suoy braza ꝑ modo che llo
conven p̃ter la spada. E poi de grãde forze gle
faro dereda lo scolar che me dredo aquesto zogo
mostra como la spada dello zugadore e intẽra posta

ꝑ la presa dello scolar che me denanzi dvi a fata
la spada intera te caduda. Tu lo pov sentir
A far forte te posso fare sonza mentir.

Questo zogo se fa ꝑ tal modo, ze che uno nudicũ
lo collo mezano contra lo mezano de parte vu
sia. e subito dada cũ coverta a le strette. E buti la sua spada a lo collo dello compagno. piglãdo la sua man dritta cũ la sua stancha de sui
stesso come aqui dipento. Butarlo ꝑ ĩ terra
senza falimento metendo lo suo pe dritto dredo
lo suo dritto.

Questa e una altra presa de butar uno ĩ
terra cũ tutta la spada. zoe che aquesto scolar incroşa cum lo zugadore della parte dritta
e passa ale strette, e cum la mane sua stancha
prende lo cubito dritto del zugadore. E subito gle butta la spada al collo piglando la
sua propria spada almezo. Ello suo dritto
pe dredo lo suo dritto dello zugadore. A cosi lo butta interra cum pocho honore.

Questo e uno altro modo de butar uno in terra. E si fa p tal modo lo scolar se incrosa cum lo zugadore de la parte dritta e si uene ale strette. E cum la mane stancha piglia la spada delo zugadore passando la mezaneta della spada e subito butta la sua spada tera e quella del zugadore propia gle mette al collo pigliando lo mantenir al mezo zoe i mezo de le mane del zugadore. E uzlo suo pe dritto de redo lo suo dritto e p tale modo lo butar in tera cu la sua spada propria.

Questo e lo mezano tor de spada chi lo sa fir tal voltar de spada se fa a questi come se fa allo fenar. Salue che le prese non sono iguali. Io prumo tor de spada liga trambedui gli brazzi. Io non uno auere quisti imparzi. Io sepero uno brazo e le mane una dall altra. El no e si forte che me la possa tegnir chel no gle la faza delle mane cadere. Come e ditto desoura Io son lo tore de spada mezano che mille uolte la fitto fior furlano.

Questo e uno altro tor de spada chiamado sotta no p tal modo se fa aquesta como fa lo soprano e sotano zoe cum tal voltar de spada. p lo camino de le altre aquesta vada. Cum la mane dritta cargando in una uolta tonda cum lo mantenir. E la mane stancha la uolta tonda debia seguir.

Questo e uno altro tor de spada che acosi fatto. Che quando uno e ale strette introsido lo scolar de meter la sua mane dritta p sotto la sua desi instesso. E piglar quella del zugadore quisi al mezo o ben erto. E subito lasar la sua spada andare in tera. E cu la man stancha de piglare la spada dello zugadore e dar da a man dritta. E uueza la sua spada p sta uolta toglandoge la spada sotto lo pomo la sp gle la uolta ton bito lo zugadore e Ello scolar cum meza p ferir lo zugado re.

15ᵛA *Questo e uno altro modo de butar uno in terra. E si fa per tal modo lo scolar se incrosa cum lo zugadore dela parte dritta, e si vene ale strette. E cum la mane stancha pigla la spada delo zugadore passando la mezamento della spada, e subito butta la sua spada in terra, e quella del zugadore propia, gle mette al collo piglando lo mantenir al mezo zoe in mezo de la mane del zugadore. E cum lo suo pe dritto dredo lo suo dritto, e per tal modo lo butar in terra cum la sua spada propria.*

This is another method of throwing someone to the ground, and it's done like this: the student crosses with the player on the right side and comes to the tight [play]. He grasps the player's sword with his left hand (passing the middle of the sword), and he immediately throws his sword to the ground and thrusts the player's own sword around his neck, grasping the hilt in the middle—that is, between the player's hands. And with his right foot behind [the player's] right, in this way he throws him to the ground with his own sword.

15ᵛB *Questo e lo mezano tor de spada, chi lo sa far tal voltar de spada se fa ad aquesto come se fa allo primar, Salvo che le prese non sono inguali.*[102] *Lo primo tore de spada liga trambedui gli brazi. Io non voio avere quigli impazi. Io sepero uno brazo e le mane una via dallaltra. El non e si forte che me la possa tegnir, chel non gle la faza delle mane cadere. Come e ditto de sovra, Io son lo tore de spada mezano, che mille volte la fatto Fior furlano.*

This is the taking of the sword from the middle: whoever knows how to do such turnings of the sword does this one just as the first is done, except that the hold isn't the same. The first taking of the sword binds both the arms, but I don't want to go to such trouble, so I separate one arm and hand from the other. No one's so strong that he could hold [his sword] and keep it from falling. As was said above, I'm the taking of the sword from the middle, which was made a thousand times by Fiore Furlano.

15ᵛC *Questo e uno altro tor de spada, chiamado sottano per tal modo se to aquesta, como fa lo soprano, e sotano zoe cum tal voltar de spada. Per lo camino de le altre aquesta vada. Cum la mane dritta cargando innanci volta tonda cum lo mantenir. E lla mane stancha la volta tonda debia seguir.*

This is another taking of the sword, called Below: do this one in the same way as you do those from above and from [the middle][103]—that is, with a turn of the sword. This one follows the method of the others, with the right hand carrying forward a full rotation with the hilt, and the left hand should follow with a full turn.

15ᵛD *Questo e uno altro tore de spada che acosi fatto. Che quando uno e ale strette incrosado lo scolar de meter la sua mane dritta per sotto la sua de si insteso. E piglar quella del zugadore quasi al mezo o ben erto. E subito lasar la sua spada andare in terra, e cum la man stancha de piglare sotto lo pomo la spada*[104] *dello zugadore e dargle la volta tonda*[105] *a man dritta. E subito lo zugadore avera*[106] *la sua spada persa. E llo scolar cum meza volta toglandoie la spada po ferir lo zugadore.*[107]

This is another taking of the sword, and it's done like this: when someone's crossed in the tight [play], the student should thrust his right hand under his [sword] and grasp the player's [sword] at the middle or above, immediately releasing his own sword to hit the ground, and with his left hand he should grasp under the pommel of the player's sword and give it a full turn to the right. Then suddenly the player will have lost his sword and the student, righting the [stolen] sword with a half turn, can strike the player.

---

[102] *se fa ad... sono inguali* partially effaced.
[103] The text says *sotano*, "from below", but appears to be referring to the previous two sword-takings.
[104] Word bisected by sword.
[105] Word bisected by sword.
[106] Word bisected by sword.
[107] Word bisected by sword.

18ʳA *No text.*

18ʳB *Io aspetto aqui in dente de zenghiar, tuti tagli e punte io so riversar, lanzi lance, e spade, e dardi chi lo sa fare, che de tuti me savero reparare. E zo che io faro, la posta de coda longa lo po fare altre tal. Io passo fora de strada alla traversa rebatendo, zo che me vene alanzado, e quando io hoe rebatuda la punta delo avanzo de larma io dubito. Che non me fiera la persona per unaltra volta me lla conviene rebater subito.*

I wait here in the Wild Boar's Tusk. I know how to turn away all cuts and thrusts, and thrown lances, and swords, and darts (from those who know how to do that), and I know how to recover from everything.

And that which I do, the Position of the Long Tail can do too: I step out of the way to the side, deflecting that which comes thrown at me. When I've deflected the point, I'm doubtful of the advance of the weapon; so that it doesn't strike at my body once again, it's best that I beat [his weapon] quickly.

18ʳC *Ad aquesto partito si e de bastone, e daga contra lanza lo scolar aspeta quello da la lanza in tuta porta di ferro voltandose la persona, e non movendo gli pie di terra nel bastone inverso parte dritta. E come quello dal lanza tra al scolaro cum sua lanza. Lo scolaro acresce lo pe stancho fora de strada rebatendo, e schivando la lanza. E passa adosso de lui senza tarda, e cum la daga lo po ferire, questo zogo se vole fare a non falire.*

For this match, which is of staff and dagger against lance, the student awaits the lancer in the Full Iron Gate, turning his body but not moving his feet from the ground nor his staff from the right side. And as the lancer attacks the student with his lance, the student advances his left foot out of the way, deflecting and dodging the lance. And he steps close to [the lancer] without delay and he can strike him with his dagger. He wants to make this play without fail.

18ʳD *Aqui comenza spada e daga a zugare, lo avantazo si e grande a chi lo sa fare. Lo magistro spetta in questa guardia cum la daga. E lla guardia se chiama dente de zenghiaro vegna tagli e punte che di quelle me so guardiar lo pe dretto cum lo rebatere indredo lo faro tornar. lo zogo stretto so a mente e non lo posso falar. A uno a uno vegna chi contra mi vole fare, che se ello non me fuzi. Io lo guastaro in uno voltare.*

Here the sword and the dagger begin to play; whoever knows how to do it has a great advantage.

The master waits in this guard with the dagger, and the guard is called Wild Boar's Tusk. Come cuts or thrusts, I know how to defend myself from these: I'll deflect as I withdraw my right foot backward. I understand the tight play so well that I can't possibly fail. Come one by one, whoever wants to work against me, and if you don't flee from me, I'll wreck you in one turn.

Jo aspetto aqui in dente de zenghiar tuti tagli e punte Io so riuerssar lanzi lance e spade e dardi chi lo sa fare che de tuti mo sauero reparare. E zo ch'io sono la posta de coda longa lo po stare altre tal. Io passo fora de strada alla trauersa rebatendo zo che me uene alanzado e quando io l'ho rebatuda la punta delo auanzo de larma io dubito. Che non me fiera la persona per un altra volta mella conuiene rebatere subito.

A da questo partito sie de bastone e daga Ʒ lanza. lo scolar aspeta quello da la lanza in tuta porta di ferro uoltandosse la persona e no mouendo gli pie di terra nel bastone in uerso parte dritta. E come quello dalla lanza tra al scolaro cum sua lanza. Lo scolaro acreste lo pe stancho fora de strada rebatendo e schiuando la lanza. E passa adosso de lui senza tarda e cum la daga lo po ferire. questo zogo se uole fare a no falire.

Aqui comenza spada e daga azugare lo auantazo sio guadagno a chi lo sa fare. Lo magistro speta in questa guardia cu la daga. Ella guardia se chiama dente de zenghiaro vegnia tagli e punte che di quelle me so guardar lo pe drieto cum lo rebatere indredo lo faro tornar lo zogo stretto so amente e non lo posso falar. Auno a uno vegnia chi contra mi uole fare che se ello no me fuzi. Jo lo guastero in uno uoltare.

Io mio magistro contra la punta fa tal coverta e subito feri in lo volto overo in lo petto. E çi daga contra spada sempre vole zogo stretto. E piu son stretto che posso ben ferire. Ouer non tu lo crenem sofire

Che lo zugadore che me denançi avesse sapudo fare tal difesa. Sello avesse la mane stacha al scolaro posta aquesto modo dredo lo suo cubito voltandolo per tal modo che aqui se monstra. Ame non bisognava fare contrario dello magistro che sta cum la daga imposta.

De allo magistro che sta imposta cum la daga qu la spada gli ho tratto de fendente per la testa. E lui passa inanzi e questa avista ello fa presta la daga gli volta ponzando lo cubito. E aquello po far ben subito. Ancora la spada cum la lezere gli po ligare. per quello modo che lo quarto zogo de spada duna mano sa fare e ancora in la daga allo terzo zogo trovarai aquella ligadura mezana. Che apsolo volto sta forada a una spanna.

Questo e uno partito che de dagi contra la spada. Aquello che ha daga e tene aquello de la spada per lo cavezo dise io te feriro cum mia daga inanzi che tu cum la spada dela guaina. Quello dela spada dise pur tra che sono aparichiado. E ài quello dela spada fa segondo che dipento aqui de dredo.

 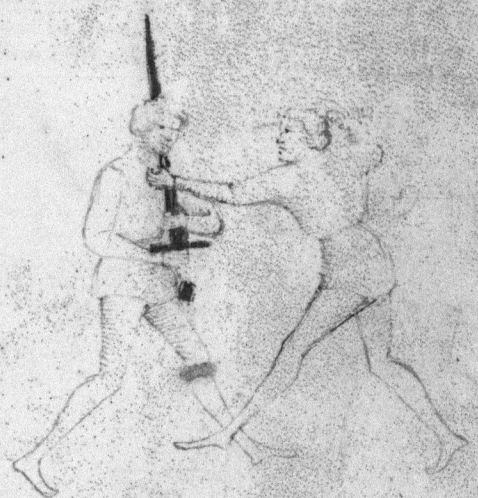

18ᵛA *Lo mio magistro contra la punta fa tal coverta e subito fieri in lo volto overo in lo petto. E cum daga, contra spada, sempre vole zogo stretto. Aqui son stretto e ti posso ben ferire. O voiᵒ non tu lo convene sofrire.*

My master makes this cover against the thrust and quickly strikes at his face or at his chest. And with dagger against sword, I always seek the tight play. Here, I'm in the tight [play] and I can strike you well. Like it or not, you must suffer.

18ᵛB *Se lo zugadore che me denanci avesse sapindo fare tal defesa. Sello avesse la mane stancha al scolaro posta aquesto modo dredo lo suo cubito voltandolo per tal modo che aqui se monstra. A me non bisognava far contrario dello magistro che sta cum la daga in posta.*

If the previous player had known how to make this defense, he would've put his left hand on the student (behind his elbow) like this, turning him in the way that's demonstrated here. Then I shouldn't have needed to do the counter to the master who stands in guard with the dagger.

18ᵛC *Se allo magistro che sta in posta cum la daga contra la spada gli vene trato de fendente per la testa. Ello passa innanzi, e questa coverta ello fa presta e dagli volta penzando lo cubito. E aquello po ferir ben subitto. Anchora la spada cum lo brazo gli po ligare,¹⁰⁸ per quello modo che lo quarto zogo de spada duna mano sa fare, e anchora in la daga allo terzo zogo troverai, aquella ligadura mezana. Che apresso lo volto sta serada a una spanna.*

If someone comes attacking with a downward blow at the head of the master (who stands in guard with the dagger against the sword), he steps forward and he makes this cover readily, and from the turn [the master] pushes his elbow and then he can strike [the player] quickly.

He can also bind the sword arm the way the fourth play of the sword in one hand does; you can find it in the third play of the dagger, too: the Middle Bind (where the hold is a hand's breadth from his face).¹⁰⁹

18ᵛD *Questo e uno partito che de daga contra la spada Aquello che ha daga e tene aquello de la spada per lo cavezo dise io te ferirò cum mia daga innanzi che tu cavi la spada dela guagina. Quello della spada dise pur tra che sono aparechiado. E cum quello dela spada fa segondo che dipento aqui de dredo.*

This is a match of dagger against sword. The one who has a dagger and holds the swordsman by the chest says, "I'll strike with my dagger before you draw your sword from the scabbard." The swordsman says, "Attack, I'm ready." And with that, the swordsman does that which is drawn hereafter.

---

¹⁰⁸ Corrected from *u* to *a*.
¹⁰⁹ Neither of these plays are present in the current manuscript.

The Flower of Battle: MS M 383

17ʳA *Quando costui leva lo brazo per darme della daga. Subito io gli o posta la da guagina apozada allo brazo suo della daga per modo che non mi po far impazo. E subito io sguaino la mia spada, e si lo posso fer inanzi chello me possa tochar cum sua daga. Anchora poria torgle la daga dela mano per lo modo che fa lo primo magistro de daga. Anchora poria ligarlo in la ligadura mezana, che lo tezo[!] zogo della daga, dello primo magistro che remedio.*

When he lifts his arm to give it to me with the dagger, I quickly put my scabbard on his dagger arm in such a way that he can't give me trouble. And I draw my sword quickly and can strike before he can touch me with his dagger.

I could take the dagger from his hand the way that the first master of dagger does, too. I could also bind him in the Middle Bind, which is the third play of the dagger (of the first master who is remedy).[110]

---

[110] Neither of these plays are present in the current manuscript.

Quando costui leva lo brazo p[er] dar me della daga.
Subito io gli o posta la daga guagina a pozada allo
brazo suo della daga p[er] modo che no[n] me po far i[m]-
pazo. E subito io squarro la mia spada e si
lo posso ferir inanzi ch'ello me possa tochar cu[n]
sua daga. Anchora p[er]zo torzele la daga de la
mano p[er] lo modo che fa lo p[ri]mo magistro de daga.
Anchora p[er]zo li p[ar]io i[n] la ligadura mezana
che lo terzo zogo della daga dello p[ri]mo magistro che remedio.

Nu' semo tri zugadori che volemo ferir questo magistro. Uno gli è di trar de punta, l'altro de taglo, l'altro vole lanzare la sua spada contra lo ditto magistro. Sì che bene serà grande fatto ch'ello non sia morto questo magistro. Che dio lo faza ben tristo.

Voi sete cativi e di questa arte sane pocho. Fate gli fatti che parole non ha logo. Vegna a uno a uno chi sa fare e po. Che se voi fosse cento tuti ve guastarò per questa guardia che così bona e forte. Io acresco lo pè che denanzi uno pocho fora de strada. E cu lo stancho io passo alla traversa. E cu questo passar io me coruzo rebatendo le spade ne torno discoverti e de ferire ve faro certi. E sì la za o spada che me ven lanzada tute lo rebatto come io ditto passando fora de strada. Segondo che voi vederete gli mie zoghi de dredo. De guadagno ve prego. E pur cum spada a una man farò mia arte.

Quello che ha ditto lo magistro io lo ben fatto. Zoè uno passar fora de strada fazando bona coverta. E lo zugadore trovo discoverto. Sì che una punta gle volo metere in lo volto per certo. E cu la man stancha volo provare. Se la tua spada posso in terra far andare.

In tuto to tronado discoverto e in la testa to ferido per certo. E sì o cum lo mio pè de dredo voio innanzi passar a più zoghi stritti porzia conti ferire. Zoè in ligadure roture, e in abrazare.

17ᵛA[111] *Noij semo tri zugadori che volemo ferir questo magistro. Uno gle di trare de punta laltro de taglo, laltro vole lanzare la sua spada contra lo ditto magistro. Si che bene sera grande fatto chello non sia morto questo magistro. Che dio lo faza ben tristo.*

We are three players who want to strike this master, one with the point, another the edge, and another wants to throw his sword against said master. It'll be a great feat indeed if this master isn't killed. May God make him suffer.

17ᵛB *Voi sete cativi e di questa arte save pocho. Fate gli fatti che parole non ha logo, vegna a uno a uno chi sa fare e po. Che se voi fosse cento tuti ve guastaro per questa guardia che cosi bona e forte. Io acresco lo pe che denanci uno pocho fora de strada. E cum lo stancho io passo alla traversa. E cum quelo passar io me covro rebatendo le spade, ve trovo discoverti, e de ferire ve faro certi. E si lanza o spada che me ven lanzada, tute le rebatto come io ditto passando fora de strada, segondo che vui vederite gli mie zoghi de dredo, de guardagli chio ven prego, e pur cum spada a una man faro mia* arte

You are wicked and you know little of this art; the things you do defy all description. Come one by one, whoever knows what they're doing and is capable, and even if you were a hundred, I would wreck all of you with this guard (which is so good and strong). I advance my forward foot slightly out of the way, and I step to the side with my left. I cover myself during that step, deflecting your swords and finding you uncovered, and I'll be certain to strike you. And if lance or sword is thrown at me, I'll deflect them all just as I've said, stepping out of the way according to that which you see in my plays hereafter. Watch what I show to you, and I'll make my art with the sword in one hand.

17ᵛC *Quello che ha ditto lo magistro io lo ben fatto, zoe chio passai fora de strada fazando bona coverta. E lo zugadore trovo discoverto. Si che una punta gle voio metere in lo volto per certo. E cum la man stancha voio provare. Se la tua spada posso in terra fare andare.*

I've done that which the master has said well—that is, I stepped out of the way, making a good cover, and I found the player uncovered such that I certainly want to thrust my point in his face. And with my left hand, I want to try to see if I can make your sword hit the ground.

17ᵛD *In tuto to trovado discoverto, e in la testa to ferido per certo. E sio cum lo mio pe de dredo voio innanzi passar asai zoghi stretti poria contra ti fare, zoe in ligadure roture, e in abrazare.*

I've found you completely uncovered, and I've struck you in the head for sure. And if I want to step forward with my rear foot, I can make many tight plays against you (that is, binds, breaks, and grappling).

---

[111] For a discussion of anomalies on 17ᵛ and 19ʳ, see the appendix by LECCESE.

The Flower of Battle: MS M 383

19ʳA *Questo e uno zogo che vole essere armato chi vole metere tal punta. Quando uno te ~~d~~ tra de punta, o de taio, tu fai la coverta, e subito metegli questa per lo modo che dipento.*

This is a play in which he who wants to thrust the point wants to be armored. When someone throws his point (or edge) at you, make the cover and quickly thrust this in the way that's drawn.

Questo ze uno zogo che uole essere armato chi
uole metere tal punta. Quando uno te d tra de
punta, o de tuio, tu fai la couerta, e subito metegli questa p lo modo che diporto.

19ᵛ *No text.*

20ʳ *No text.*

20ᵛ   *No text.*

The Flower of Battle: MS M 383

Appendix: Riddled in Ink: A Stylistic Comparison of Artwork in MS M 383 and Novati's Facsimile

by Jay Leccese

Proposition

This article proposes that Francesco Novati's 1902 facsimile reproduction of *Flos Duellatorum* contains clear stylistic discrepancies that can both elucidate its connection to the original Pisani Dossi manuscript, and identify connections between the Pisani Dossi and other manuscripts of *The Flower of Battle*. These inconsistencies surface in the bottom two images of Novati Carta 13^A, and in all four images of Novati Carta 13^B. This article seeks to illustrate that all six of these images not only differ stylistically from the rest of the artwork in the Novati facsimile, but indeed share strong stylistic similarities with the artwork of the Morgan. The primary thesis of this article argues that the illustrations identified in the Novati facsimile on Carta 13^{AB} were in fact drawn by the Morgan artist. Secondarily, this article seeks to present evidence that the Novati facsimile is an exact reproduction of the Pisani Dossi manuscript.

Relevance

This proposition is relevant because it validates the accuracy of the Novati facsimile, provides new evidence for understanding the order of construction of the Morgan and Pisani Dossi manuscripts, suggests that the Pisani Dossi artist had direct access to illustrated source material at the time of the manuscript's construction, and underlines the possibility of artistic duplication within the corpus of *The Flower of Battle*.

Methodology

This article relies upon detailed side-by-side image comparison of artwork and forensic details from both the Novati facsimile and the Morgan manuscript. It begins with a stylistic analysis in which artwork samples from the Morgan manuscript are compared with the images in question on Novati 13. Both sets of images are then compared with corresponding artwork samples taken from elsewhere in the Novati facsimile. The analysis of the artwork takes into consideration the execution of the following elements: overall presentation, head, torso, upper limbs, lower limbs, clothing, shading, and contour. The stylistic analysis will conclude by presenting an inconsistent image from the Morgan manuscript for close comparison with the artwork of the Novati facsimile. Subsequently, a forensic analysis will be conducted to examine evidence which may elucidate the relationship of the Novati facsimile to the original Pisani Dossi manuscript, as well as provide insights on the order in which the Pisani Dossi and Morgan manuscripts were constructed. This analysis will begin by examining anomalies pertaining to gold and silver foil application, and will investigate the significance of drawn margins, ghost images, follicle marks, and handwriting comparisons.

Stylistic Analysis

Those who have studied the Novati facsimile closely may have already recognized that there is something disparate about the illustrations on the two pages depicted below. Something about them feels different. This sensibility may be difficult to articulate, other than to say that the images are somehow not like the other drawings in the Novati. Those who have also closely studied the artwork of the Morgan manuscript may make the additional observation that these illustrations seem not only different, but familiar. It was the intersection of these two realizations that generated the flashpoint for the research leading up to the current analysis. The following investigation has been structured to allow the reader to make both of these observations simultaneously with each case.

Overall Presentation

These images have been provided so that the reader may have a general impression of how the artistic styles in question differ when viewed side by side. The reader should note that the black ink of the Novati facsimile may not accurately represent the original color of the Pisani Dossi manuscript.

Noy colpi mezani andamo traversando
Dal zenochio in su andamo guastando
E refudemo le punte fora de strada
E rezdopiando lo colpo defesa e derada
E si noy del mezano colpo itamo i sedel
Asay cu tali colpi guastamo zent

Ponte semo de grandissima offensione
E a tuti colpi fazemo questione
Venenose semo piu ch spente
E piu ch tuti colpi alcidono zete
E noy ponte ali colpi si disemo
Tanto no travzet ch noy avemo

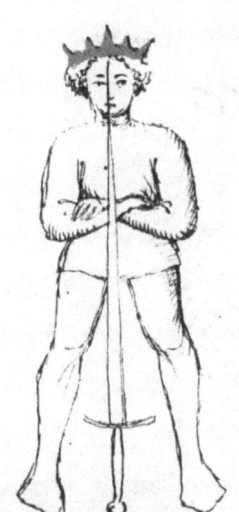

Planture de spada e trare tayo e punta
La guardia che io ho mete me moti
Vegna auno auno chi contra mi uole far
Che cu tuti io uoio contrastar
E chi uole uedere couerte e fizire
Tor de spada e ligadure senza falir
Guardi ghi mie scolari como san fare
De elli non trouan zirario no ano pare

Cu passo o fato couerta cu mia spada
E aquella in lo peto subito te itudei

[Carta 13 A.]

P ferirte anchora cu questa mia piuta
La man sinistra ala spada t'o zunta

Aqui io t'o ferido in la tua testa
P la coverta ch'io fata acosi fata

P la mane ch'io posta sotto tuo elzo
Si tua spada no va i tera di me gitezo

P tal modo te distrovo p ferirte de piuta
P uendegarme de ti d'ogni iguria conta

APPENDIX: RIDDLED IN INK

Morgan 18ʳ (D), Novati 13ᴬ (C), Novati 16ᴮ (A)

Morgan 16ᵛ (B), Novati 13ᴬ (D), Novati 14ᴬ (B)

Morgan 15ᵛ (C), Novati 13ᴮ (B), Novati 14ᴬ (A)

It is notable in looking at these images that the figures in the examples from Novati 13ᴬᴮ appear stilted and delicately drawn, like the images from the Morgan. In contrast, the examples of standard Novati artwork, shown in each case on the right, appear rough and heavy-limbed by comparison. This general impression will be validated when these elements are more closely examined.

The Flower of Battle: MS M 383

Head Comparison

The images below provide examples of heads that are representative of the artistic styles in question.

Morgan 14ᵛ (A), Morgan 14ʳ (B), Morgan 13ᵛ (B), Morgan 14ᵛ (C)

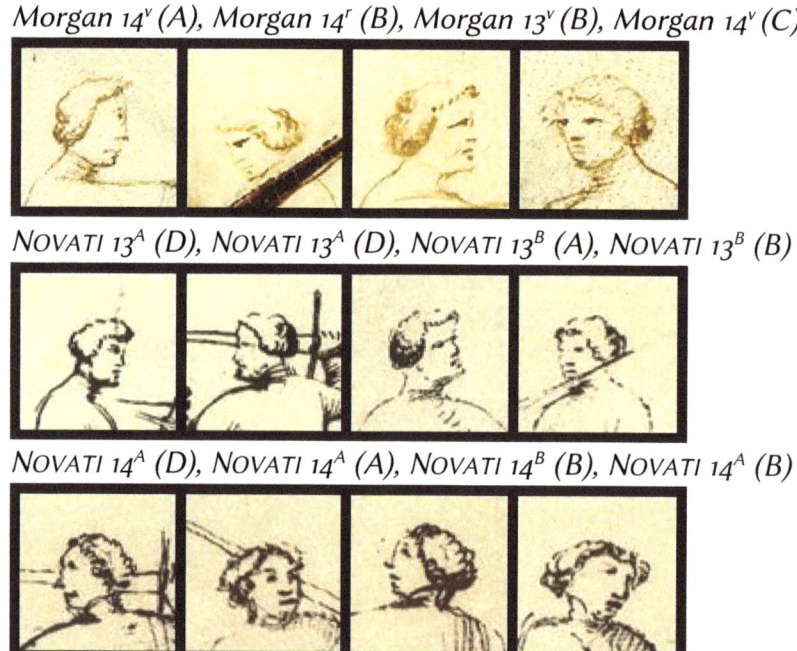

Novati 13^A (D), Novati 13^A (D), Novati 13^B (A), Novati 13^B (B)

Novati 14^A (D), Novati 14^A (A), Novati 14^B (B), Novati 14^A (B)

Note the contour of the hairline as depicted in the Morgan artwork. The hairline turns abruptly downward at the corner of the forehead, and then curves backward behind the neck. This shape of the hairline is repeated in the examples from NOVATI 13^AB. The examples of standard NOVATI artwork taken from 14^A and 14^B, on the other hand, do not follow this pattern. Seen in profile, the hairline tends to slope straight back from the front of the forehead. When seen partially from the front, it tends to be parted acutely in the middle. Likewise the treatment of the faces on NOVATI 13^AB resemble that of the Morgan rather than the typical faces of the NOVATI.

APPENDIX: RIDDLED IN INK

Torso Shape Comparison

The images below provide examples of torsos that are representative of the artistic styles in question. Note the overall shape of the torsos, especially in profile, with close attention to the contour of the chest and back.

Morgan 15ᵛ (B), Morgan 15ᵛ (C), Morgan 16ʳ (C), Morgan 16ʳ (B)

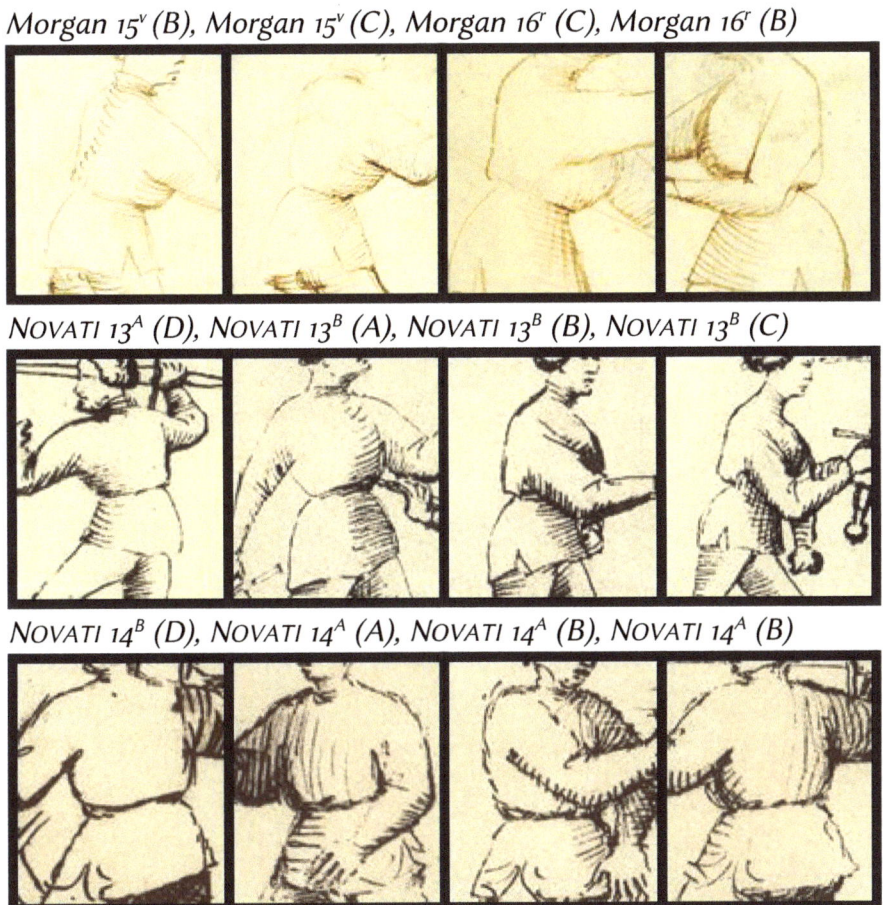

Novati 13ᴬ (D), Novati 13ᴮ (A), Novati 13ᴮ (B), Novati 13ᴮ (C)

Novati 14ᴮ (D), Novati 14ᴬ (A), Novati 14ᴬ (B), Novati 14ᴬ (B)

Note the relative narrowness of the waist in NOVATI 13ᴬ (D) and NOVATI 13ᴮ (A). The hourglass shape of the torso is similar to the style of the torsos in the Morgan, shown immediately above. Additionally, note the shape of the upper torsos of NOVATI 13ᴮ (B) and NOVATI 13ᴮ (C). In both of these images, the chest bulges out with exaggerated roundness, whereas the contour of the lower back juts out abruptly from the waistline and then turns upward to form a right angle. This is characteristic of the Morgan artwork and is seen throughout the latter. In contrast, the torsos of typical figures in the NOVATI have a less pronounced hourglass shape due to a more gradual taper at the waist, most significantly, lack the awkward bulges at the back and chest.

The Flower of Battle: MS M 383

Upper Limb Comparison

The images below provide examples of upper limbs that are representative of the artistic styles in question. Note the proportions.

Morgan 13ᵛ (C), Morgan 15ᵛ (B), Morgan 16ʳ (C), Morgan 18ᵛ (B)

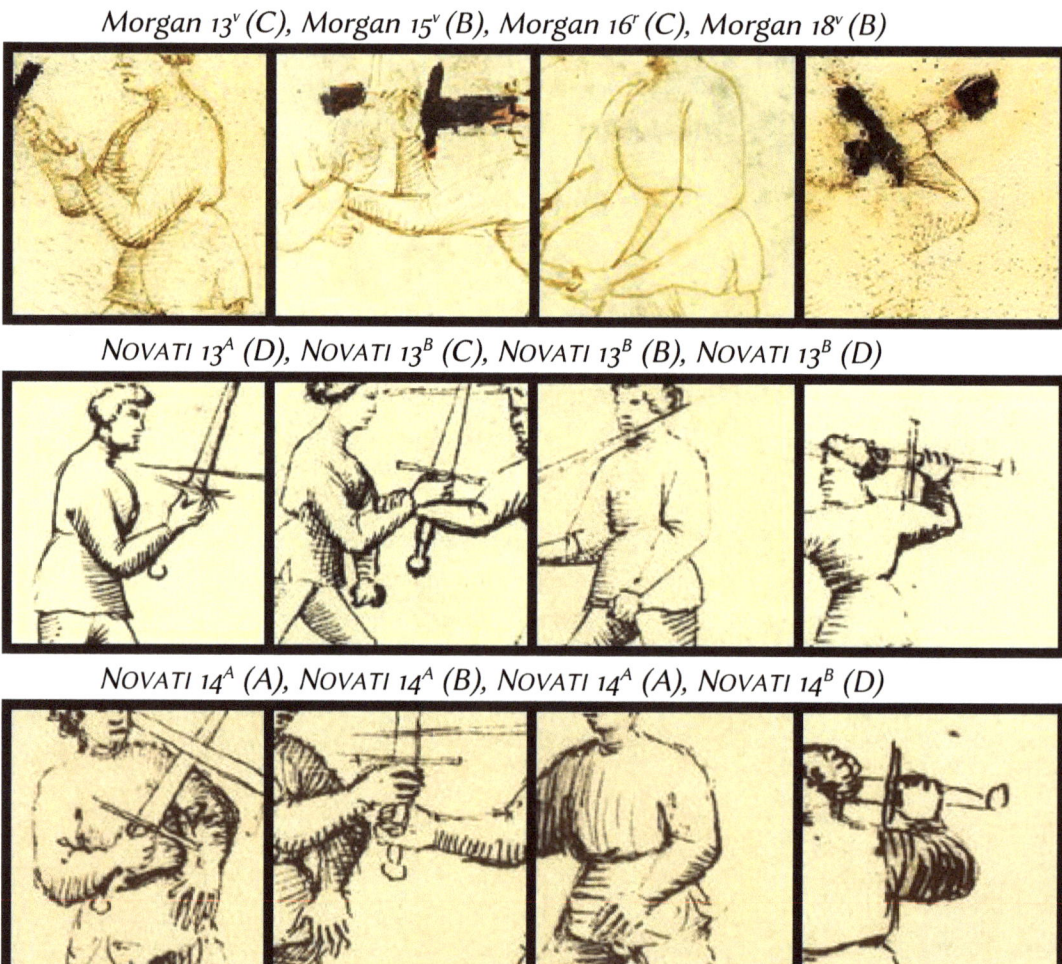

Novati 13ᴬ (D), Novati 13ᴮ (C), Novati 13ᴮ (B), Novati 13ᴮ (D)

Novati 14ᴬ (A), Novati 14ᴬ (B), Novati 14ᴬ (A), Novati 14ᴮ (D)

Note the taper of the arms from shoulder to elbow, and elbow to wrist, in the images from Novati 13ᴬᴮ. Like the Morgan images, these taper more gradually, and tend to be thinner at the elbow and at the wrist, as compared with the characteristic Novati images below them. The arms typical of Novati artwork tend to be broader, especially at the upper forearm, often giving them the appearance of being stunted or shrouded in billowing sleeves. Also of note is the size of the hands in the Novati images shown at the bottom. These are strangely large and are often drawn with splayed fingers. The images from Novati 13ᴬᴮ have much smaller hands that tend to be closed in a fist, like those of the Morgan.

APPENDIX: RIDDLED IN INK

Lower Limb Comparison

The images below provide examples of lower limbs that are representative of the artistic styles in question. The execution of the legs is one of the most striking signs of stylistic variation within the NOVATI. Note the proportions.

Morgan 9ᵛ (D), Morgan 13ᵛ (B), Morgan 15ᵛ (D), Morgan 15ᵛ (B)

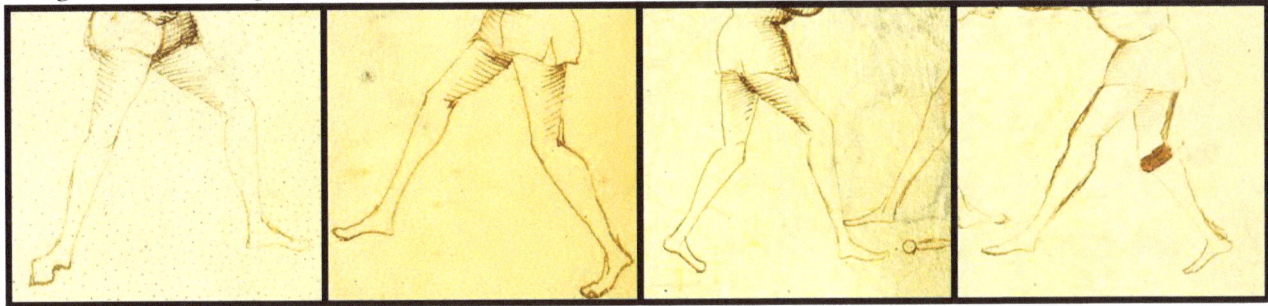

NOVATI 13ᴬ (C), NOVATI 13ᴮ (B), NOVATI 13ᴮ (C), NOVATI 13ᴮ (C)

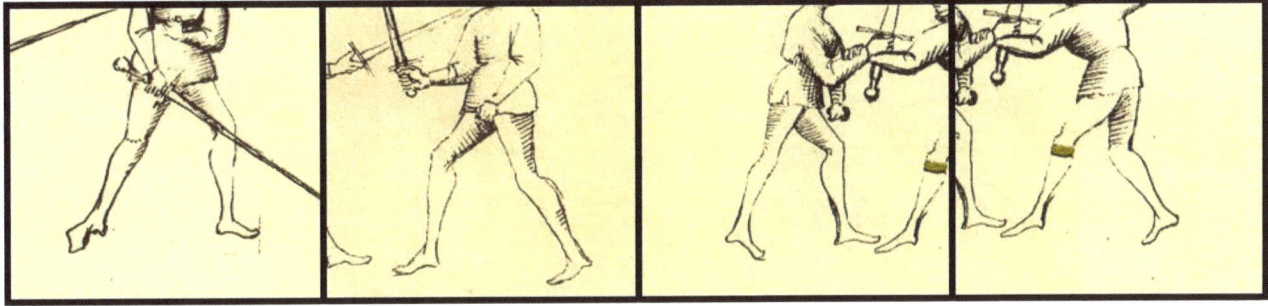

NOVATI 14ᴬ (C), NOVATI 14ᴬ (A), NOVATI 14ᴬ (B), NOVATI 14ᴬ (D)

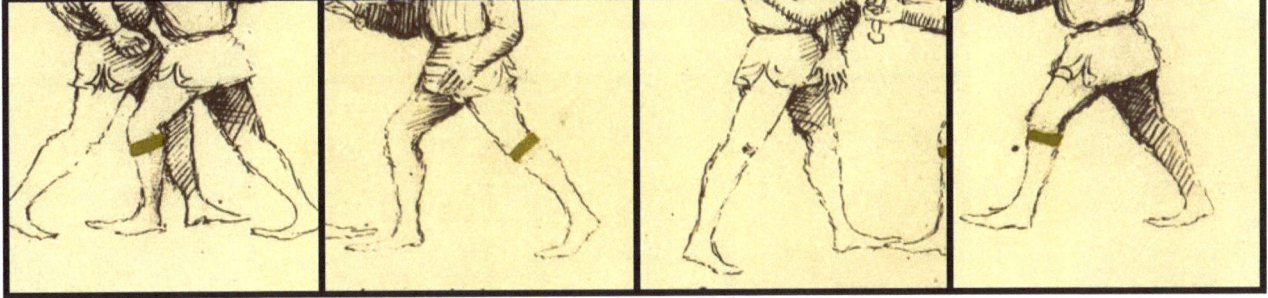

Note how narrow and elongated the legs are in NOVATI 13ᴬᴮ. The lower legs are exceptionally long and narrow with prominent calves, while the upper legs are comparatively short and thin above the knee. This gives the legs a stilted and delicate appearance. This is distinctive of the Morgan artwork. In contrast, the legs typical of the NOVATI appear thick and heavy, with broader ankles and thighs. Also telling are the differences in the treatment of the feet. Like those in the Morgan artwork, the feet in NOVATI 13ᴬᴮ are short in comparison to the proportion of the legs, and typically rounded at the toe.

The Flower of Battle: MS M 383

The typical NOVATI feet differ in that they are relatively elongated, with a very narrow taper toward the toe that often ends in a point. Lastly and most notably, the oddly hoof-like foot in NOVATI 13ᴬ (C) is strikingly similar to the foot in the image above it, taken from the Morgan. This hoof-like drawing of the foot is a hallmark of the Morgan artist, and is seen numerous times in that manuscript.[112]

Clothing Comparison

The images below provide examples of clothing details that are representative of the artistic styles in question. Readers familiar with the Morgan will recognize the clothing details below as being among the few adornments of that manuscript.[113]

Morgan 9ʳ (A), Morgan 9ᵛ (A), Morgan 16ᵛ (B), Morgan 15ᵛ (B)

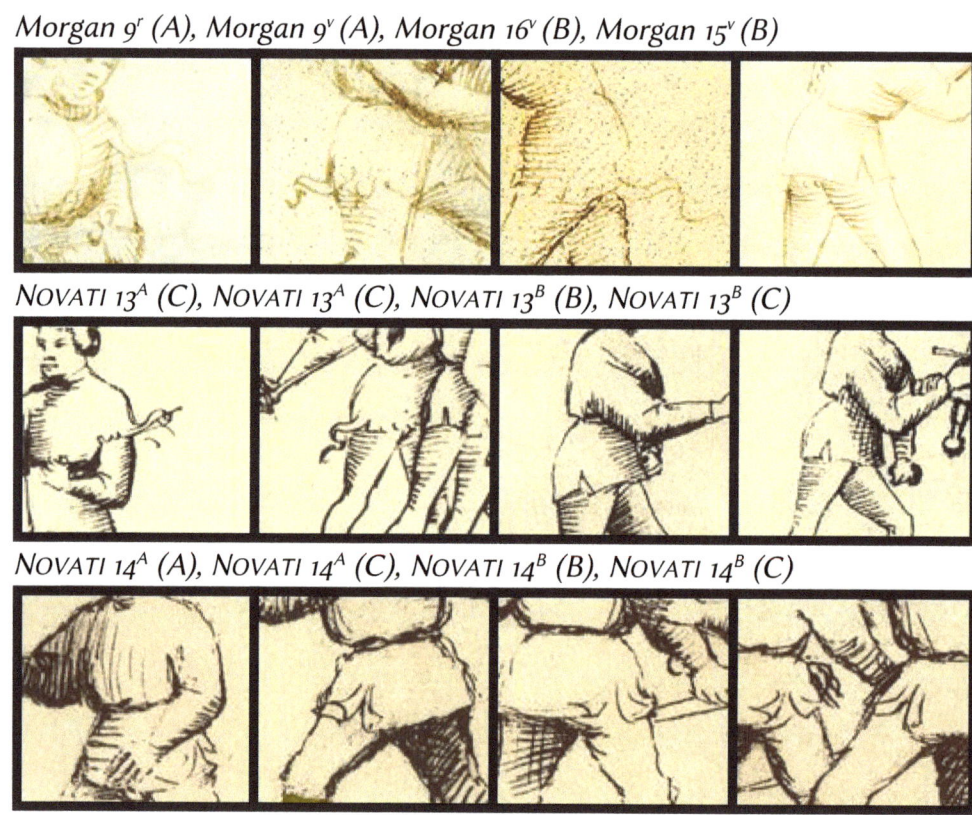

NOVATI 13ᴬ (C), NOVATI 13ᴬ (C), NOVATI 13ᴮ (B), NOVATI 13ᴮ (C)

NOVATI 14ᴬ (A), NOVATI 14ᴬ (C), NOVATI 14ᴮ (B), NOVATI 14ᴮ (C)

Note the flowing upper sleeve in NOVATI 13ᴬ (C), as well as the fluttering hem of the figure's jacket to the right, also from NOVATI 13ᴬ (C). This detail is identical to several which appear throughout the Morgan manuscript, some of which have been displayed directly above. This type of detail does not appear anywhere else in the NOVATI. Also note the V shaped cut in the

---

[112] Morgan, fol. 9ʳ, 9ᵛ, 10ᵛ, 12ʳ, 13ʳ, 14ᵛ.
[113] Morgan, fol. 3ʳ-9ᵛ, 11ᵛ, 16ᵛ.

coat hem of the figures in NOVATI 13^B (B) and 13^B (C). This simple cut is typical of the Morgan artwork, displayed above. Standard NOVATI artwork, by contrast, generally shows triangular sections of fabric folding outward in either direction from this notch. Some of the figures illustrated in the Paris also display this stylistic treatment of the coat hem with two triangular flaps of fabric turning outward from the notch. It does not appear in the Morgan.

Shading Comparison

The images below provide examples of shading that are representative of the artistic styles in question. The reader is directed especially to the variations in density and uniformity of the crosshatching, as well as the general shape of the shaded areas.

Morgan 10^v (C), Morgan 11^r (A), Morgan 14^r (D), Morgan 16^r (B)

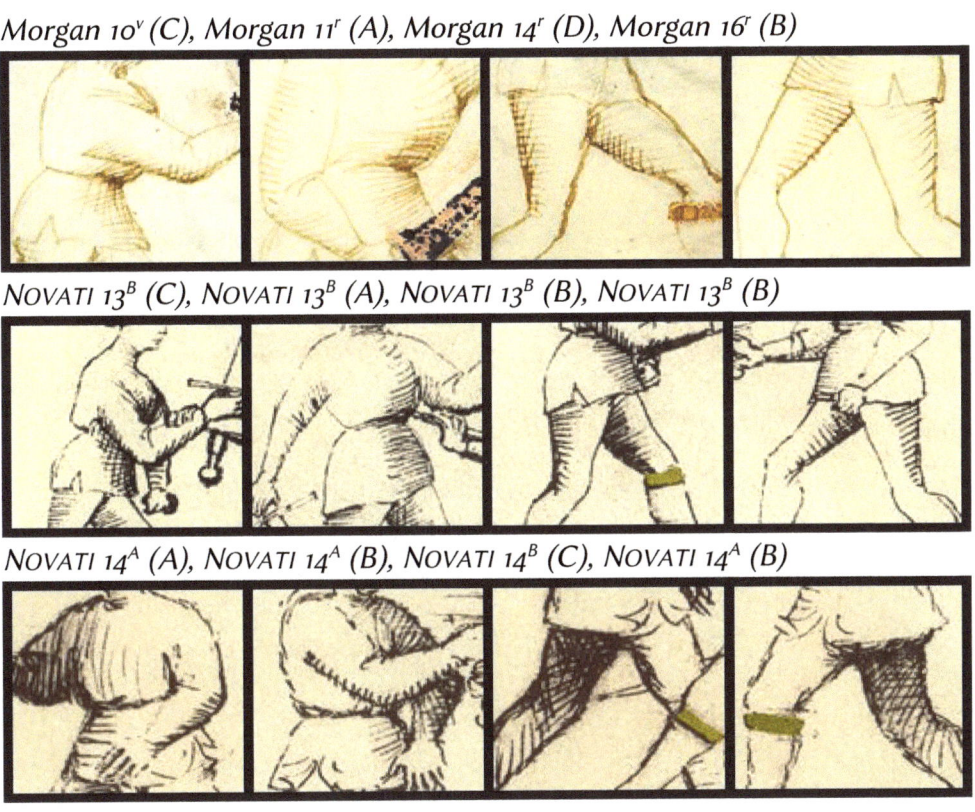

NOVATI 13^B (C), NOVATI 13^B (A), NOVATI 13^B (B), NOVATI 13^B (B)

NOVATI 14^A (A), NOVATI 14^A (B), NOVATI 14^B (C), NOVATI 14^A (B)

Note the regular spacing of the hatching on the torsos from NOVATI 13^B. This shading is drawn sometimes with simple horizontal lines, and other times with intersecting crosshatching; these tend to be delicate and fairly evenly spaced. This approach to shading is consistent with that of the Morgan. By contrast, the crosshatching in typical NOVATI artwork tends to be heavier and more irregular. This same pattern extends to the treatment of the arms and legs. Often, the shading in the NOVATI completely darkens an entire upper arm or leg with heavy strokes. In contrast, the shading of the limbs in

NOVATI 13^AB is done more conservatively. For instance, shading on the arms consists of short tick marks on the lower or inner edge of the arms, like that in the Morgan. Most notably, the cross hatching on the legs of the figures in NOVATI 13^AB consist almost exclusively of small triangular sections on the backs or insides of the thighs, tapering toward the knee. This is a very distinct characteristic of the Morgan artwork.

Contour Comparison

The images below provide examples of contours that are representative of the artistic styles in question. The reader is directed to note especially the number of pen strokes in a given contour.

Morgan 16ʳ (C), Morgan 16ʳ (B), Morgan 14ʳ (A), Morgan 15ᵛ (A)

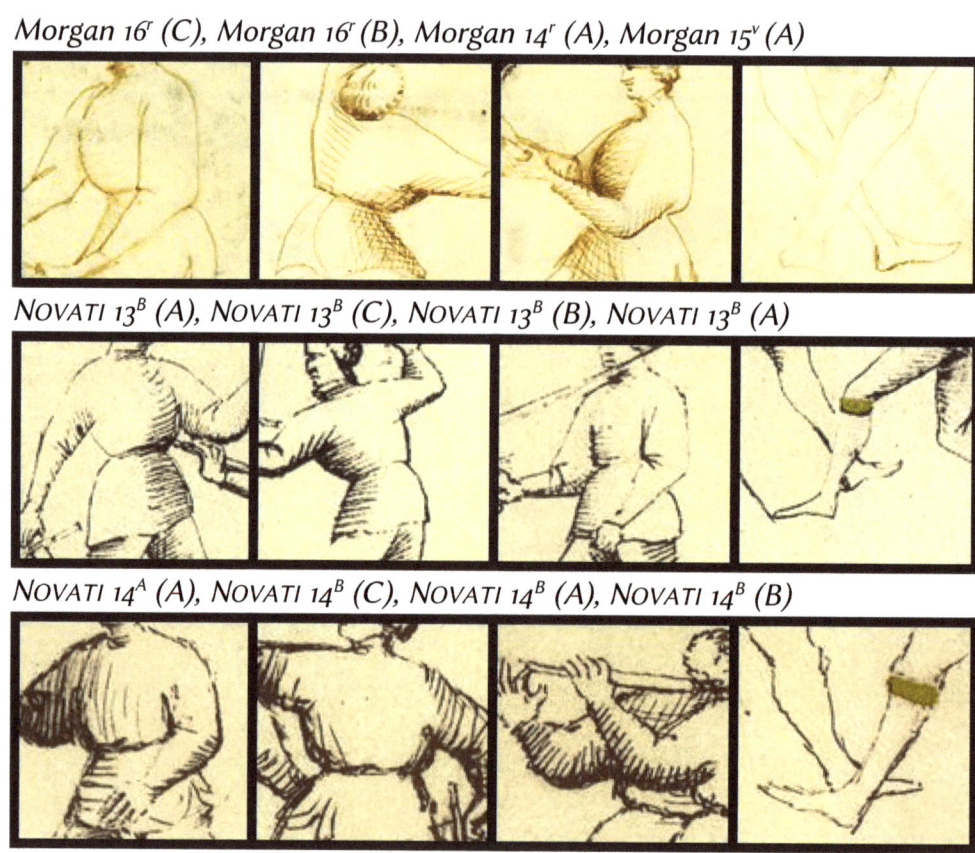

NOVATI 13^B (A), NOVATI 13^B (C), NOVATI 13^B (B), NOVATI 13^B (A)

NOVATI 14^A (A), NOVATI 14^B (C), NOVATI 14^B (A), NOVATI 14^B (B)

Note the relative smoothness and continuity of the outer contours in the images from NOVATI 13^B. Like those of the Morgan artwork above, the lines tend to be drawn very economically, using a small number of continuous pen strokes. Their execution suggests a controlled and steady hand. By contrast, the contours of the standard NOVATI artwork shown at the bottom are rough and choppy. Each contour appears to have been executed with the effort of many short pen strokes that often crisscross and overlap. If there is a single fundamental element which on the one hand separates the drawings

APPENDIX: RIDDLED IN INK

of Novati 13[AB] from the rest of the Novati artwork, and on the other ties it closely with the artwork of the Morgan, is the execution of the lines themselves. This distinction is categorical, and will be found upon every examination of the images in question. More even than style, it is the mark of a different hand.

A Retouched Image in the Morgan

Here the analysis turns to consider the curious case of Morgan 10ʳ (D), illustrating the guard *porta di ferro mezana* alongside other *poste* for fighting with sword in armor.[114] This image stands out from the other images on the page by virtue of its conspicuously bold ink. It is one of several instances throughout the Morgan where an artist has partially inked over some faded images—a common practice sometimes done long after the original construction.[115] While it is the dark ink that initially draws attention, it is the style of the image which holds the greatest curiosity. After having closely surveyed the artistic styles of both the Morgan and Novati, one may see that in this case the places have been reversed. This single image bears nearly every stylistic element of the Novati which had been previously identified in contrast to the Morgan artwork. If it can be said that Morgan 10ʳ (D) was at some point redrawn, it can also be said with confidence that it was done so by none other than the artist of the Pisani Dossi. In both cases, the hand of the artist is the same. Morgan 10ʳ (D) is displayed below between examples from both the Morgan and the Novati.

Morgan 10ʳ (A), Morgan 10ʳ (D), Morgan 10ᵛ (B)

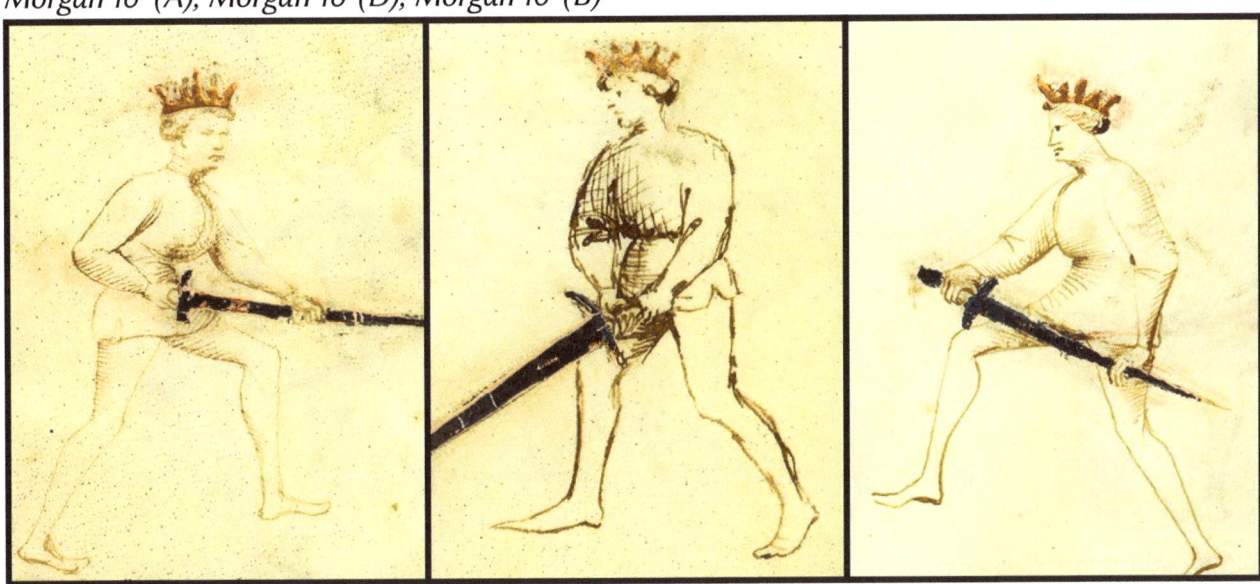

---

[114] Morgan, fol. 10ʳ.
[115] Alexander 39.

The Flower of Battle: MS M 383

Novati 12ᴬ (C), Morgan 10ʳ (D), Novati 18ᴮ (D)

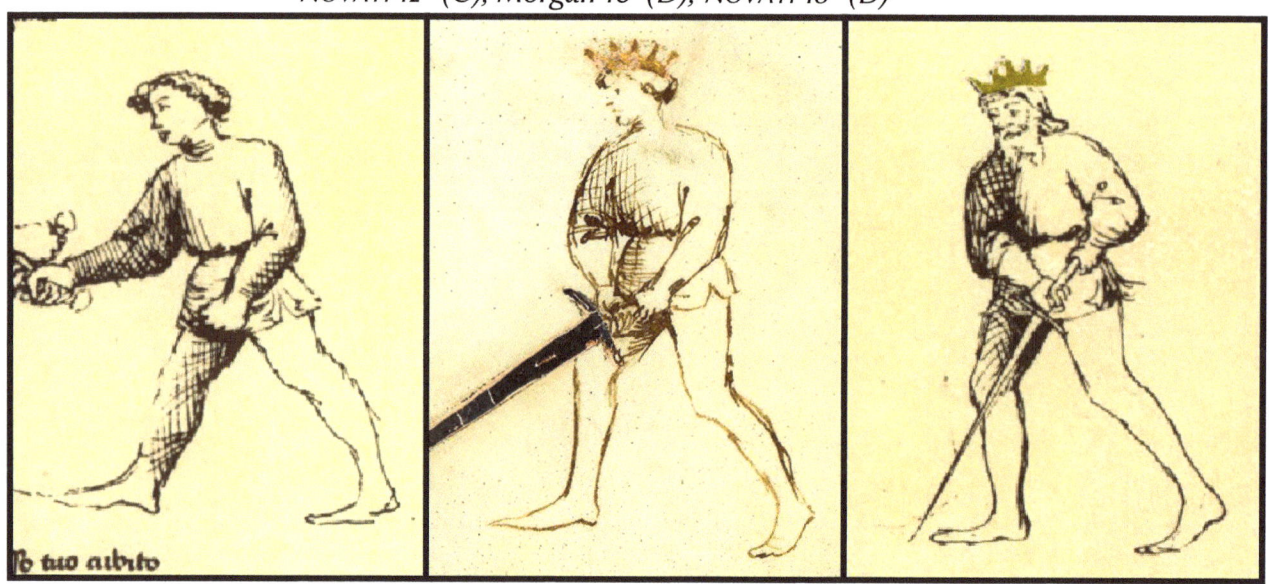

Features which set Morgan 10ʳ (D) apart from the rest of the Morgan artwork, and connect it stylistically with the Pisani Dossi artwork, include the following. Notably: the shape of the feet; the proportion and thickness of the legs, especially at the ankles, calves, and above the knee; the proportion and thickness of the arms, especially at the wrists and elbows; the height of the armpit relative to the shoulder; the shape of the torso above the waistline; the breadth of the shoulders; the treatment of the hair; the pattern of the cross-hatching; and perhaps most strikingly, the bold and sporadic roughness of the lines. Details of Morgan 10ʳ (D) alongside similar details from the Novati have been provided below, as well as a grayscale comparison of the Morgan and Novati images to correct for ink color variation.

Head

Novati 6ᴮ (F) reversed, Morgan 10ʳ (D), Novati 9ᴬ (E)

APPENDIX: RIDDLED IN INK

Torso

NOVATI 7^A (E), Morgan 10^r (D), NOVATI 8^B (A)

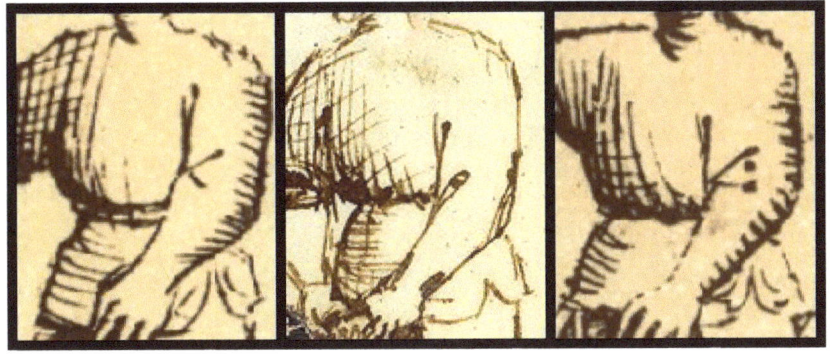

Upper Limbs

NOVATI 7^B (A) reversed, Morgan 10^r (D), NOVATI 8^A (A)

Lower Limbs

NOVATI 6^B (F), Morgan 10^r (D), NOVATI 18^B (B)

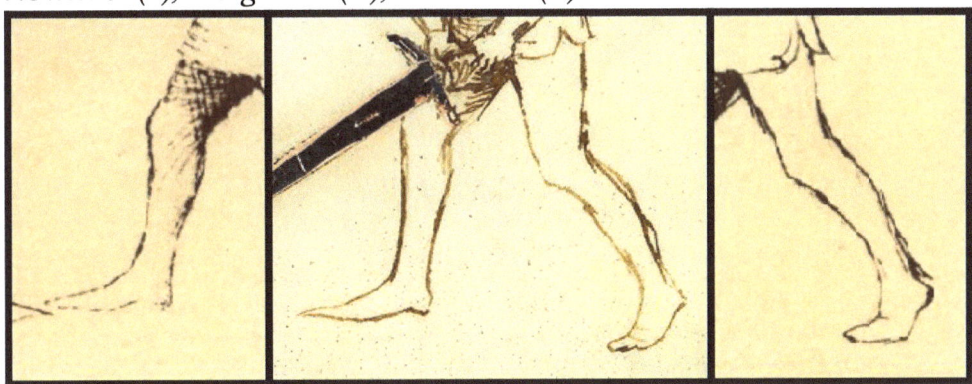

Grayscale

Morgan 12^r (D), Morgan 10^r (D), Novati 18^B (D)

The evidence presented in the stylistic analysis above supports the initial proposition of this article that the artwork on NOVATI Carta 13^{AB} is consistent with the style identified in the Morgan manuscript. It also provides evidence that Morgan 10^r (D) was drawn by the Pisani Dossi artist. At face value these observations should not be shocking, since it was common for multiple artists to collaborate on the same manuscript, even to the extent of significant differences in style.[116] However we may infer from the difference in fading of the ink that the drawing attributed here to the Pisani Dossi artist was likely executed considerably later than the other illustrations in the Morgan. If this is the case, it is also possible that the inclusion of Morgan artwork in the Pisani Dossi was not due to artistic collaboration, but was simply an inclusion of single leaves executed at an earlier date by another artist—a practice which was not uncommon in the later Middle Ages.[117] It is even possible that they were once part of a larger volume of *Fior di Battaglia* whose remnants now comprise the Morgan manuscript. This is not unlikely considering that the Morgan appears to be missing many pages, among which are the very plays that are shown on NOVATI Carta 13^{AB}.[118]

Taken together, these propositions strongly suggest that the artist and compiler of the Pisani Dossi had direct access to the pages now contained

---

[116] ALEXANDER 47, 95.

[117] Ibid, p 36.

[118] While these plays are indeed represented in the Morgan, the illustrations on folio 17^v are clearly drawn by a different artist. They closely match the corresponding images in Getty fol. 20^{rv}, and may have been drawn by that artist or a very diligent copyist. The same is the case with the image on Morgan 19^r, which closely corresponds to the image on Getty fol. 21^v, with the inclusion of many armor elements from that manuscript.

in the Morgan manuscript. The following section will examine further evidence that pertains to the accuracy of the Novati facsimile, the degree to which the artist of the Pisani Dossi manuscript had access to the Morgan manuscript, and their relative order of construction.

Forensic Analysis

This section examines the relevant forensic evidence which has arisen in investigation of Novati Carta 13^{AB}, as well as Morgan Folio 10^r (D). It also examines evidence which supports the contention that the Novati facsimile is a close reproduction of the Pisani Dossi manuscript, as well as evidence which may elucidate the order of construction of the Morgan and Pisani Dossi manuscripts.

Gold and Silver Leaf

The Flower of Battle is noteworthy for its use of a system of gilded crowns and garters in order to more clearly structure its pedagogy. This feature is visible in all known copies of the treatise, and seems to be unique to the teachings of Fiore dei Liberi. While examples of drawn crowns surface in *Die Blume des Kampfes*, specifically in the Erlangen, they are rare decorations which seem to have lost all pedagogical significance.[119] In addition to this well-known system of gilded crowns and garters, the Morgan manuscript is unique amongst the corpus of *The Flower of Battle* in that it also identifies masters, scholars, and counters by the application of silver leaf to their weapons. This silver leaf has turned black with oxidation, and is perhaps the most striking and recognizable feature of the Morgan. Shown below are examples of gold and silver leaf application from the Morgan manuscript, and examples of gold leaf application from the Novati facsimile. Note that the drawings from the latter, which have been identified as Morgan artwork, have no silver leaf.

---

[119] Erlangen, fol. 21^r, 31^v, 33^r.

Morgan 16ʳ (B), Morgan 16ʳ (C)

NOVATI 13^A (D), NOVATI 13^B (B)

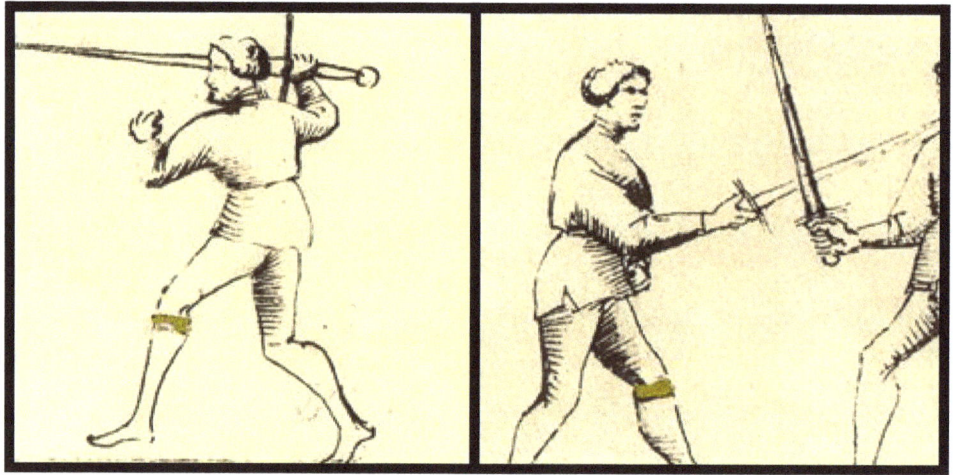

After surveying the stylistic evidence in the images of NOVATI Carta 13^AB, the analysis has indicated that they were almost certainly drawn by the Morgan artist. As has already been noted, these may have been single leaves which were eventually captioned and gathered into the Pisani Dossi manuscript, or may have even been drawn from an earlier compilation whose remnants now survive in the Morgan. If this exchanging of pages is granted, then it is likely that the application of the silver leaf to the swords in the Morgan was done at some time after the compilation of the Pisani Dossi, for while the scholars in Carta 13^AB were clearly drawn by the hand of the Morgan artist, their swords lack the now characteristic silver leaf. Moreover, if one considers that Morgan 10ʳ (D) was drawn at a later date by the Pisani Dossi artist, then it may be reiterated that the application of gold and silver leaf came later, for in Morgan 10ʳ (D) both the gold and silver leaf appear to overlay the newer ink. This is shown in detail below.

APPENDIX: RIDDLED IN INK

Morgan 10ʳ (D)

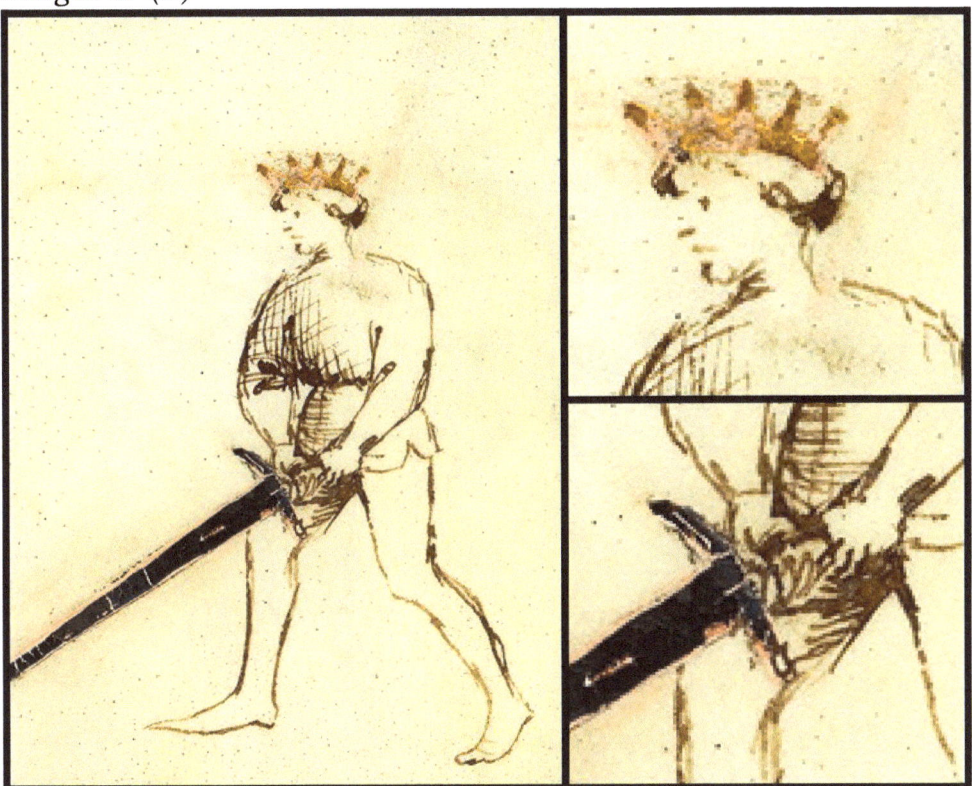

Rule Marks

The use of vertical and horizontal rule marks to align text and images is a common feature in many original manuscripts.[120] Examples of this can be seen clearly in the digital images of all three original manuscripts of *The Flower of Battle* currently available to the public (the Morgan, Getty, and Paris). These features are also identifiable in the NOVATI facsimile. Examples have been provided below.

Getty 18ʳ (A), Getty 18ʳ (B), Getty 23ᵛ (C)

Morgan 2ᵛ (A), Morgan 2ᵛ (A), Morgan 10ᵛ (A)

---

[120] ALEXANDER 40.

The Flower of Battle: MS M 383

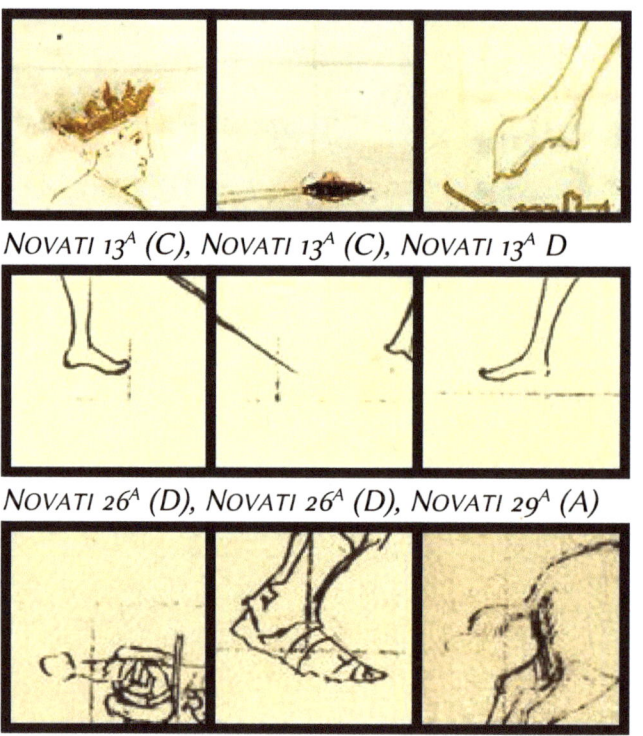

Novati 13ᴬ (C), Novati 13ᴬ (C), Novati 13ᴬ D

Novati 26ᴬ (D), Novati 26ᴬ (D), Novati 29ᴬ (A)

Novati 26ᴬ (C, D)

Note the lines drawn in Novati 13ᴬ, depicted above. These are evidence of rule marks which were originally drawn to help center the images on each quadrant of the page. Such lines are also faintly visible on many other pages of the Novati, including 15ᴮ, 19ᴬ, 20ᴮ, 21ᴬ, 22ᴬ, 22ᴮ, 26ᴬ, and 29ᴬ. These lines support the contention that the Novati facsimile is an exact reproduction of the original Pisani Dossi manuscript.

APPENDIX: RIDDLED IN INK

Ghost Images

Due to the partial translucency of vellum, images from the reverse side of a page will often be visible through the page, especially when light is shown or reflected from behind. This effect is notable in the digital images of the three manuscripts of *The Flower of Battle* currently available. Examples from the Morgan and NOVATI have been provided below.

Morgan 10ᵛ (C), Morgan 10ᵛ (C) reversed, Morgan 10ʳ (D)

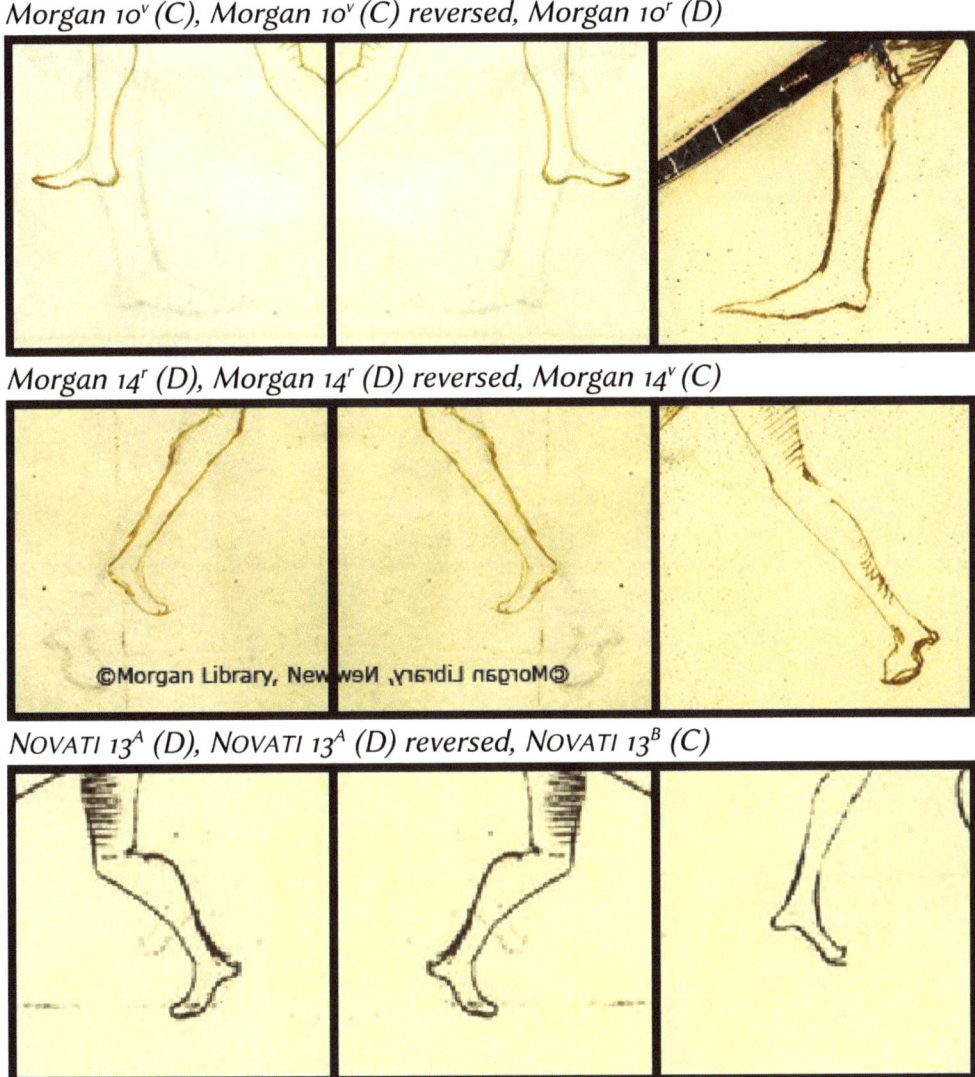

Morgan 14ʳ (D), Morgan 14ʳ (D) reversed, Morgan 14ᵛ (C)

NOVATI 13ᴬ (D), NOVATI 13ᴬ (D) reversed, NOVATI 13ᴮ (C)

Note the example above from Morgan 10ᵛ (C), where the ghost image of a large foot appears below one of the figures. When reversed, one can see that this ghost image corresponds exactly with the foot of the figure on the reverse side of Folio 10. Likewise, Morgan 14ʳ (D) contains the ghost image of a foot showing through from the reverse side of the page. When looking at NOVATI 13ᴬ (D), the same phenomenon is observable. There is the ghost image of a forefoot which, when reversed, can be matched up with the image on the other side of the page.

Novati 21ᴮ (D), Novati 21ᴮ (D) reversed, Novati 21ᴬ (C)

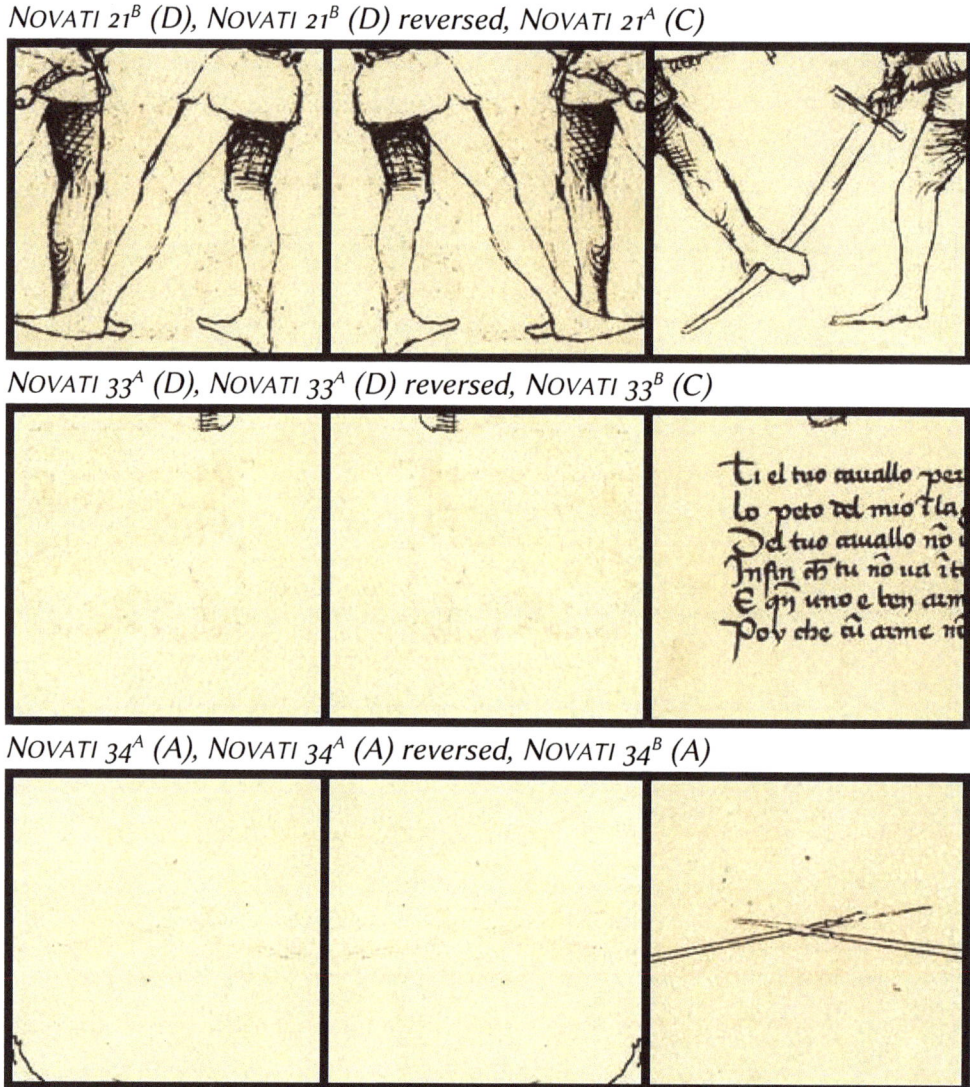

Novati 33ᴬ (D), Novati 33ᴬ (D) reversed, Novati 33ᴮ (C)

Novati 34ᴬ (A), Novati 34ᴬ (A) reversed, Novati 34ᴮ (A)

The images above provide examples of other ghost images seen throughout the Novati. Note in 21ᴮ (D) the very faint image of a bent sword. When reversed, it corresponds with the bent blade on the other side of the page. Novati 33ᴬ (D) contains a very faint ghost image of the block of text on Novati 33ᴮ (C). Lastly, in Novati 334 (A), a ghost image of crossed lances can be seen. Such ghost images appear on nearly every page, and are especially visible on the print edition, where they are in fact printed faintly in black ink onto the page. Such evidence strongly suggests that the Novati images were reproduced directly with early facsimile technology.

APPENDIX: RIDDLED IN INK

Follicle Marks

True vellum parchment is made of animal hide which has been treated, stretched, and scraped to remove hair and excess skin.[121] As such, it is known to have two sides, the smooth side being from the inside of the animal, and the coarser side from the outside of the animal. The outside, or hair-side of vellum parchment is easily distinguishable by the remnants of hair follicles, which are visible as small dark spots on the page.[122] These follicle marks are visible in the digital images of all three manuscripts currently available. Black dots resembling follicle marks are also clearly printed in ink onto the pages of the NOVATI facsimile. Examples have been provided below.

Getty 10ʳ (C), Morgan 15ʳ (A)

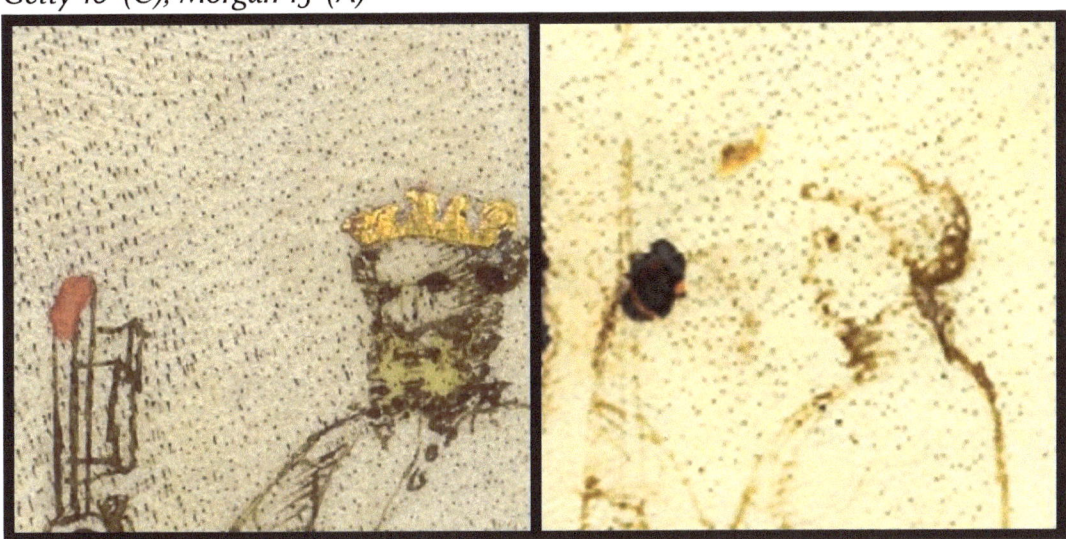

NOVATI 15ᴬ (B), NOVATI 20ᴬ (D), NOVATI 26ᴮ (D)

---

[121] JOHN MUMFORD and others. "Recording the physical features of Codex Sinaiticus". *The British Library.* http://codexsinaiticus.org/en/project/conservation_physDesc.aspx. Retrieved July 2016.
[122] Ibid.

145

The Flower of Battle: MS M 383

While it would require a codicological expert to count the spacing and number of these dots to determine with certainty that they are consistent with animal hair follicles, they are a detail which we would expect to see reproduced in a facsimile reproduction using photogravure or other photo-mechanical technology. Alongside the rule marks and ghost images, these dots—whether the reproduction of follicle marks or other residue—weigh heavily in favor of the proposal that the Novati facsimile is an exact photo-mechanical reproduction of the Pisani Dossi manuscript.

Handwriting

In returning to Novati Carta 13^A with the consideration that the page may have come directly from an earlier volume whose remnants are now contained in the Morgan manuscript, one may venture with caution that at least parts of the Morgan manuscript may originally have been without caption. This possibility becomes clear when comparing the handwriting of Novati Carta 13^A both with other writing samples from the Novati, and with samples from the Morgan: despite having Morgan artwork on the bottom of the page, Novati Carta 13^A is captioned in the hand that captioned the rest of the Pisani Dossi. Samples have been shown below.

Morgan 14^r (A), Novati 13^A (A)

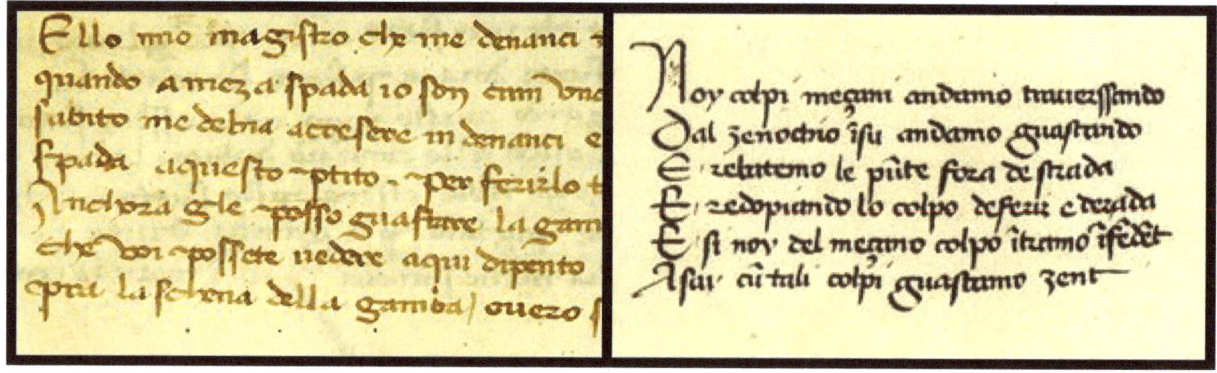

Novati 13^A (A), Novati 16^A (A)

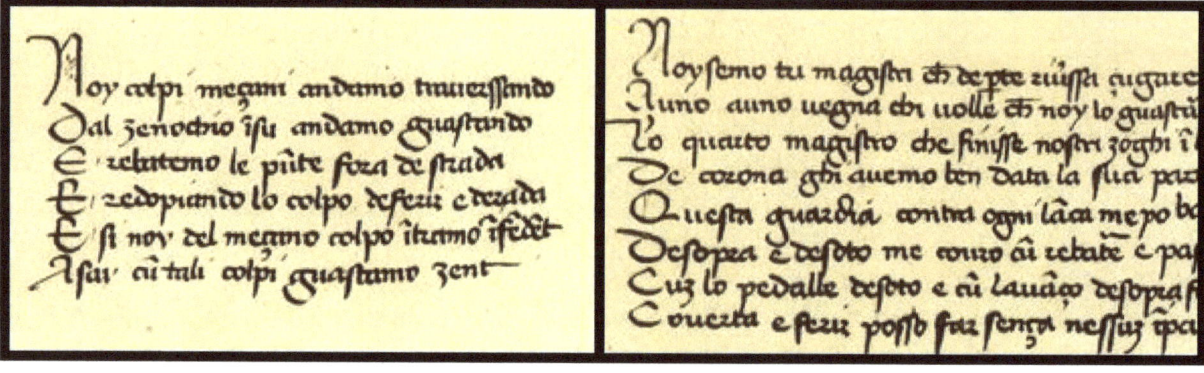

It can be seen immediately that the script on NOVATI 13^A resembles the script from NOVATI 16^A, far more than it resembles the script from Morgan 14^r. Some of the specific differences between these scripts will be examined below.

Alignment

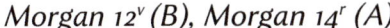
Morgan 12^v (B), Morgan 14^r (A)

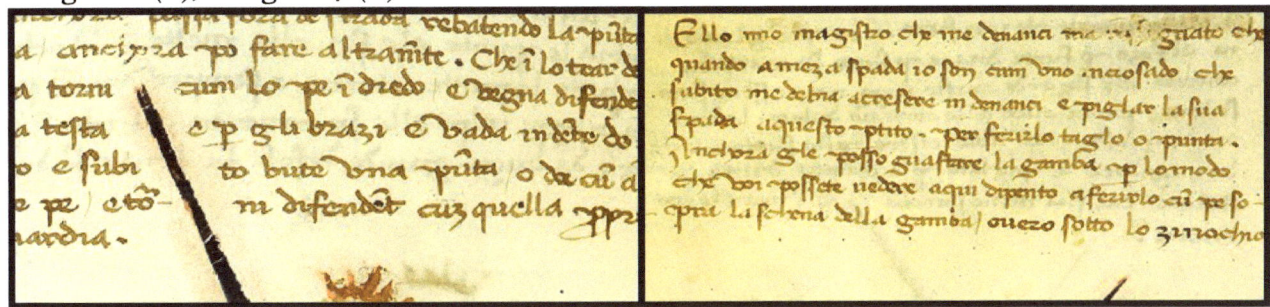

NOVATI 7^B (C), NOVATI 12^A (B)

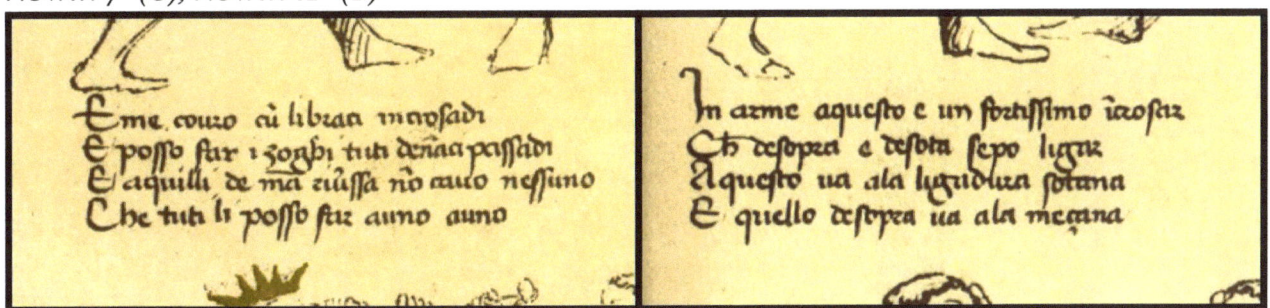

The first difference that jumps out between the respective text of the Morgan and NOVATI is the variation in precision of alignment. While the text in the Morgan appears aligned in blocks, a close examination shows that the lines of text often deviate significantly from a horizontal orientation. Further, in some cases the spacing between lines or words varies to allow for images which appear to have already been on the page at the time of writing.[123] In contrast, the text in the NOVATI tends to be more regularly spaced and evenly squared on the page. It deviates very little from a horizontal orientation and uniform spacing, even when crammed tightly between images, as seen above.

---

[123] Morgan, fol. 6^r, 7^r, 9^r, 10^v, 11^r, 12^v, 14^r, 14^v, 15^r, 15^v, 17^v, 18^r, 18^v.

Punctuation

Morgan 12ᵛ (B), Morgan 14ʳ (A), Novati 7ᴮ (C), Novati 11ᴬ (C)

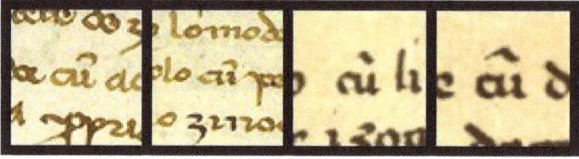

Note the execution of the abbreviation "cû" for *cum*. In the Morgan the carat appears rounded on the top, whereas in the Novati it tends to be pointed.

Morgan 8ᵛ (A), Morgan 15ᵛ (B), Novati 12ᴮ (B), Novati 13ᴬ (B)

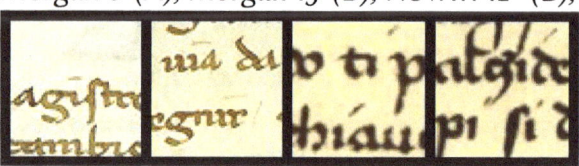

Note the execution of the punctuation on the "i". In the Morgan this accent is drawn horizontally, whereas in the Novati it is drawn ascending along a diagonal up and to the right.

Letters

Morgan 14ʳ (A), Morgan 15ᵛ (B), Novati 12ᴮ (C), Novati 13ᴬ (C)

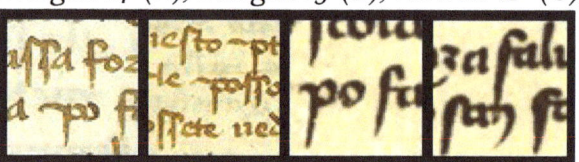

Note the execution of the descenders on the letters "f", "p", and "s". In the Morgan these descend downward roughly perpendicular to the line of the script. In the Novati, they descend at a slight angle downward and to the left. In both cases with ascenders such as on the letter "l", the same trend is followed. In the Morgan they are mostly vertical, whereas in the Novati they are written with the same angle as the descenders.

Morgan 15ᵛ (B), Morgan 17ᵛ (B), Novati 11ᴬ (C), Novati 13ᴬ (A)

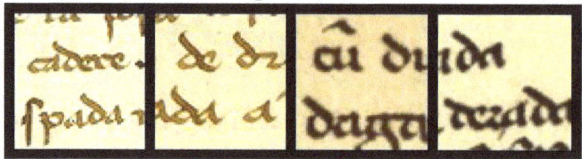

Note the execution of the letter "d". In the Morgan the ascender on the letter slants upward and to the left, with a slight upward hook at the terminal. In the Novati, the ascender on the "d" slants more strongly to the left, and the upward hook tends to be absent or less pronounced.

Morgan 14ʳ (A), Morgan 17ᵛ (B), Novati 11ᴬ (C), Novati 13ᴬ (A)

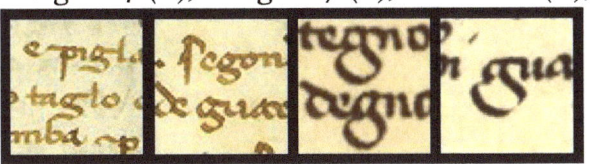

Note the execution of the letter "g". In the Morgan the lobe of the "g" forms an open hook upward that either terminates before it passes the leftmost edge of the bowl of the letter, or else terminates in a tail that swoops downward and to the left. In the Novati, the lobe of the "g" descends to form a wide loop that turns upward to connect to the bowl of the letter.

These observations have been cursory, and it is for a forensic typographer to examine them in more detail. However, they have presented significant evidence which supports the other findings of this article. On the one hand they have shown differences between the script of the Novati and the Morgan, and on the other they have solidified the connection of the writing on Novati 13ᴬᴮ with that in the rest of the Novati.

Title Page

As remarked above, the handwriting on Novati 13ᴬᴮ can be cross referenced to handwriting throughout the Novati and shown to be identical. Moreover, if this exercise is continued, one will find this handwriting consistent with that on the title page of the Novati, which is surrounded by marginal illumination. The latter shows intricacies of style and wear that strongly indicate that it is an exact photo-mechanical reproduction. On closer inspection, the handwriting and the illumination appear to be inextricably connected, for in the upper left-hand corner the letters are blurred by the same water damage that has disrupted parts of the illumination. Given that this handwriting also matches that in the rest of the Novati, it follows that the script reproduced throughout the Novati must be that of the original scribal hand. Examples are provided below.

The Flower of Battle: MS M 383

Novati 2^A

Novati 2^A, Novati 16^B (B)

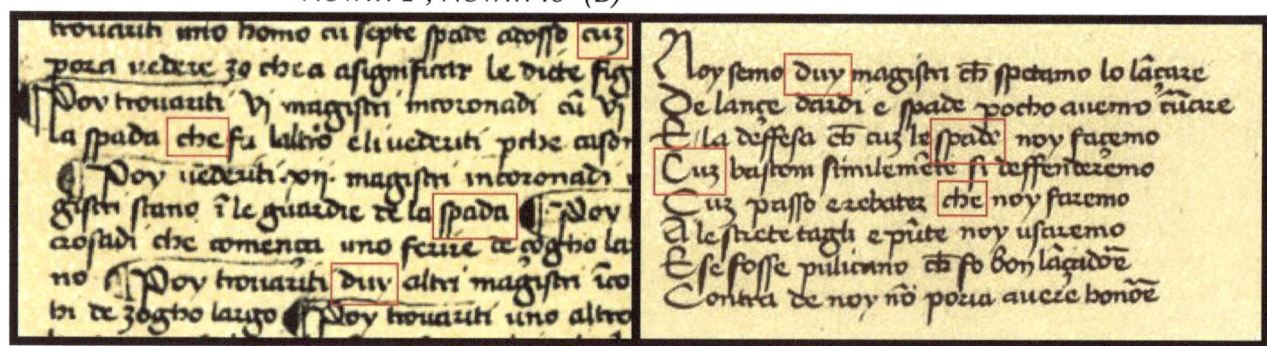

Note the line of tonal variation along the top of NOVATI 2^A, which appears to delineate the top edge of a page that has been reproduced onto the printed image. This may be a product of early photo-mechanical reproduction technology. Some type of damage, possibly due to water, is also visible in the top left corner of the page. This has caused some of the illumination to be blurred, along with some of the letters of the script. Flaking of pigment from the illuminated initial can also be seen here, possibly a result of the same event that damaged the marginal illumination. A detail from the text of NOVATI 2^A is presented above alongside text from NOVATI 16^B (B) for comparison. Compare for example the words *cuz*, *che*, *spada*, and *duy*, outlined in red. It will be found that, within the normal deviation to be expected of a single hand, the execution of the letters is in fact the same.

Discussion and Conclusion

This article began with the proposition that certain images in the NOVATI facsimile, namely those on Carta 13^AB, bore a strong stylistic resemblance to the illustrations in the Morgan manuscript. It proceeded with an analysis that presented specific examples which highlighted these resemblances. This analysis has left little room to doubt that there is a significant stylistic connection. Moreover, it strongly indicates that the Morgan artist himself illustrated the images in question. The analysis also raised the case of Morgan 10^r (D), which with perhaps even greater certainty was shown to have been illustrated by the artist of the Pisani Dossi. Subsequent evidence pertaining to the application of metallic foil and written script was presented to elucidate the likely order in which these events took place. Taken together, the evidence suggests that the artist of the Pisani Dossi had direct access to the collection of drawings whose remnants are now contained in the Morgan manuscript, during which time he illustrated an image on Morgan 10^r, may have acquired a loose leaf from that same collection, and would almost certainly have reproduced some of its content.

The evidence also raises the possibility—given that this loose leaf did not have text prior to the writing of the Pisani Dossi—that the content of the entire Morgan manuscript may once have consisted of images without captions, much like Vienna. If this is the case, then the images now in the Morgan may once have served a function similar to certain pattern books, which contained collections of model images intended for reproduction.[124] This interpretation would both account for the compositional similarity between the images of the Pisani Dossi and the Morgan, while accounting for their significant disparity of textual content. It would also imply an additional source used in the creation of the Pisani Dossi, either in the form of another manuscript, or an advisor—perhaps the author himself—who was well-versed in Fiore's Art.

With the expected release of digital images of the Pisani Dossi, it is of perhaps diminished importance that the evidence presented in this article also suggests that the NOVATI is an exact facsimile reproduction of the Pisani Dossi. Nonetheless it is likely to be proven accurate, for even if there were no other evidence, the ability to analyze illustrations in close detail and identify specific artistic styles casts strong doubt on the contention that the NOVATI images are radically different from their originals. Even if it were accepted that the NOVATI illustrations were reproduced by hand in 1902, it would have required tracing or copying of incredible diligence to retain the details which have been noted in this article.

---

[124] ALEXANDER 110-111.

In addition to stylistic observations, however, this article has presented forensic evidence that would render the possibility that the NOVATI was hand-drawn by a modern artist incredibly unlikely. The release of the Pisani Dossi images should be anticipated with this caution, for while further detail and forensic evidence may become available, it is likely that the content will not be radically different or more extensive than what FRANCESCO NOVATI has already made available. If this proves to be the case, the wordless testaments of ink and foil will have to be plumbed to their fullest depths in order to more fully understand what is left of *The Flower of Battle*.

Suggestions for Further Research

The conclusions of this article, if granted, would open up innumerable doorways for further research. One such avenue which is currently being investigated is the possibility of artistic duplication across the extant manuscripts of *The Flower of Battle*. Such research would profit immensely from a greater understanding of the order of construction of each manuscript, and the extent of the source material available to their respective artists. This article has sought to elucidate precisely these relationships.

About the Author

Jay Leccese is an independent researcher focusing on close study of the illustrations in the four extant manuscripts of *The Flower of Battle*, as well as the three manuscripts collectively referred to as "Die Blume des Kampfes".

Bibliography

Primary sources (with abbreviations)

Manuscripts

Erlangen
    Erlangen, Universitätsbibliothek Erlangen-Nürnberg, ms. B.26
Getty
    Los Angleles, J. Paul Getty Museum, ms. Ludwig XV 13
Morgan
    New York, Morgan Library and Museum, ms. M.383
Paris
    Paris, Bibliothèque nationale de France, ms. Latin 11269
Pisani Dossi
    Corbetta, Museo archeologico Villa Pisani Dossi, [manuscript]
Rome
    Rome, Biblioteca Nazionale Centrale, MS Vitt.Em.1324
Vienna
    Vienna, Österreichische Nationalbibliothek, Cod. 5278
Zeno
    San Daniele del Friuli, Biblioteca Guarneriana, ms. XXIV

Incunables

Pozzo
    del Pozzo, Paride. *De duello, vel De re militari in singulari certamine*. Napoli: Sixtus Reissinger, 1472.

Secondary sources

DEACON
    DEACON, JACOB HENRY. "Prologues, Poetry, Prose and Portrayals: The Purposes of Fifteenth Century Fight Books According to the Diplomatic Evidence". *Acta Periodica Duellatorum* 4(2): 69-90. 2016. doi:10.36950/apd-2016-014
HATCHER
    HATCHER, COLIN. *The Flower of Battle: MS Ludwig XV13*. Ed. TRACY MELLOW. Tyrant Industries, 2017.
HAYES
    HAYES, SEAN. "Memory and Performance: Visual and Rhetorical Strategies of Il Fior di Battaglia". *Can These Bones Come to Life? Insights from Reconstruction, Reenactment, and Re-creation* 1:62-69. Wheaton, IL: Freelance Academy Press, 2014.

LEONI
: LEONI, TOM. *Fiore de' Liberi's Fior di Battaglia*. 1st ed. Lulu.com, 2009. 2nd ed. Wheaton, IL: Freelance Academy Press, 2012.

LEONI AND MELE
: LEONI, TOM AND GREG MELE. *Flowers of Battle, Volume 1: Historical Overview and the Getty Manuscript*. Wheaton, IL: Freelance Academy Press, 2017.

LODÀ
: LODÀ, FRANCESCO (in Italian). *Florius. De arte luctandi. Traduzione e cura del testo di Francesco Lodà*. Rome: Bonanno Editore, 2014.

MALIPIERO
: MALIPIERO, MASSIMO. *Il Fior di battaglia di Fiore dei Liberi da Cividale: Il Codice Ludwing XV 13 del J. Paul Getty Museum*. Udine: Ribis, 2006.

MARSDEN
: MARSDEN, RICHARD. *The Flower of Battle: MS Latin 11269*. Trans. BENJAMIN WINNICK. Tyrant Industries, 2019.

MONDSCHEIN 2011A
: MONDSCHEIN, KEN. "Notes on Bibliothèque Nationale MS Lat. 11269, *Florius de Arte Luctandi*". *Arms & Armour* 8(2):117-22. October 2011. doi:10.1179/174962611X13097916223163

MONDSCHEIN 2011B
: MONDSCHEIN, KEN. *The Knightly Art of Battle*. Los Angeles: J. Paul Getty Museum, 2011.

MONDSCHEIN 2018
: MONDSCHEIN, KEN. "On the Art of Fighting: A Humanist Translation of Fiore dei Liberi's Flower of Battle Owned by Leonello D'Este". *Acta Periodica Duellatorum* 6(1): 99-135. 2018. doi:10.2478/apd-2018-0004

MONDSCHEIN AND MELE
: MONDSCHEIN, KEN AND GREG MELE. *Flowers of Battle, Volume 3: Florius de Arte Luctandi*. Wheaton, IL: Freelance Academy Press, 2017.

NOVATI
: NOVATI, FRANCESCO (in Italian). *Flos Duellatorum, Il Fior di Battaglia di Maestro Fiore dei Liberi da Premariacco*. Bergamo: Instituto Italiano d'Arte Grafiche, 1902.

ПАЛІЙ
: ПАЛІЙ, ЛЮДМИЛА ІВАНІВНА (in Ukrainian). "Витоки художнього стилю ілюстрацій «Квітки Битви» майстра Фіоре Фріулано деі Ліберí". *Науковий вісник Ужгородського університету* 2(43):84-102, 2020. doi:10.24144/2523-4498.2(43).2020.217366

PORZIO AND MELE
: PORZIO, LUCA AND GREG MELE. *Arte Gladiatoria Dimicandi: 15th*

Century Swordsmanship of Master Filippo Vadi. Union City, CA: Chivalry Bookshelf, 2002.

RAPISARDI
RAPISARDI, GIOVANNI (in Italian). *Fiore de' Liberi Flos Duellatorum - in armis, sine armis equester et pedesta*. Gladitoria Press, 1998.

RUBBOLI and CESARI 2002
RUBBOLI, MARCO and LUCA CESARI (in Italian). *Flos Duellatorum. Manuale di Arte del Combattimento del XV secolo*. Rome: Il Cerchio Iniziative Editoriali, 2002.

RUBBOLI and CESARI 2005
RUBBOLI, MARCO and LUCA CESARI (in Italian). *L'arte Cavalleresca del Combattimento*. Rome: Il Cerchio Iniziative Editoriali, 2005.

WINDSOR 2013
WINDSOR, GUY. *Veni Vadi Vici. A Transcription, Translation and Commentary of Philippo Vadi's De Arte Gladiatoria Dimicandi*. Spada Press, 2013.

WINDSOR 2018
WINDSOR, GUY. *The Art of Sword Fighting in Earnest: Philippo Vadi's De Arte Gladiatoria Dimicandi*. Spada Press, 2018.

WINDSOR 2020
WINDSOR, GUY. *From Medieval Manuscript to Modern Practice: The Longsword Techniques of Fiore dei Liberi*. Spada Press, 2020.

ZANUTTO
ZANUTTO, LUIGI (in Italian). *Fiore di Premariacco ed I Ludi e Le Feste Marziali e Civili in Friuli*. Udine: D. Del Bianco, 1907.

Other works cited

ALEXANDER
ALEXANDER, JONATHAN J.G., *Medieval Illuminators and Their Methods of Work*. New Haven and London: Yale University Press, 1992.

CLEMENS AND GRAHAM
CLEMENS, RAYMOND and TIMOTHY GRAHAM. *Introduction to Manuscript Studies*. Ithaca, NY: Cornell University Press, 2007.

HUGHES
HUGHES, STEVEN. "Soldiers and Gentlemen: The Rise of the Duel in Renaissance Italy". *Journal of Medieval Military History* 5: 99-152. Boydell & Brewer, 2007.

MELLINI
MELLINI, GIAN LORENZO. "Disegni di Altichiero e della sua scuola, 2." *Critica d'Arte* LIII-LIV:9-19, 1962.

About the Author

Michael Chidester is the Editor-in-Chief of Wiktenauer and, as Director of the Wiktenauer, an officer of the non-profit HEMA Alliance.

Michael has been studying historical European martial arts since 2001. He was a member of the Association for Renaissance Martial Arts until 2006, where he achieved the rank of general Free Scholar, and he acted as the ARMA Provo Study Group Leader from 2007 until its dissolution in 2009. Michael co-founded the True Edge Academy of Swordsmanship in 2009, and until late 2010 was senior instructor at its Provo, Utah branch.

In 2012, Michael was appointed to the newly-established position of Director of the Wiktenauer by the HEMA Alliance general council, formalizing the role of principal designer and editor that he had assumed in early 2010. In 2013, these efforts earned him a HEMA Scholar Award for Best Supporting Researcher.

Michael has lectured at the *Historical Swordplay Symposium* at the Massachusetts Center for Interdisciplinary Renaissance Studies (including offering the keynote in 2014), *Life, the Universe, & Everything: the Marion K. "Doc" Smith Symposium* at Brigham Young University, and numerous HEMA events in Europe and America including *Blood on the River, Broken Point, Fechtschule America, Fechtschule New York, HEMAG Dijon*, the *Iron Gate Exhibition, Longpoint, Meyer Symposium*, the *Purpleheart Armory Open, Swordsquatch*, and the *Western Martial Arts Workshop*.

He has authored or edited various books, including *The Flower of Battle: MS M.383, "The Foundation and Core of all the Arts of Fencing": The Long Sword Gloss of GNM Manuscript 3227a* (2020), *The Illustrated Meyer* (2019), *The Recital of the Chivalric Art of Fencing of the Grand Master Johannes Liechtenauer* (2015), and *The Flower of Battle of Master Fiore de'i Liberi* (2015).

In 2010, Michael received a Bachelor of Arts in Philosophy from Brigham Young University, with minor degrees in Logic and Military Science and additional coursework in Italian and Spanish. He developed a certain fluency in the latter while living abroad in Mexico from 2002 to 2004. Michael is a Research Scholar of the Meyer Freifechter Guild, a founding member of the Society for Historical European Martial Arts Studies (SHEMAS), a member of the Western Martial Arts Coalition (WMAC), and a Lifetime Member of the HEMA Alliance.

www.ingramcontent.com/pod-product-compliance
Lightning Source LLC
Chambersburg PA
CBHW040800240426
43673CB00015B/407